Science and Spiritu͟͟͟͟͟͟͟͟͟

Until the end of the eighteenth century, almost everyone believed that the empirical world of science could produce evidence for a wise and loving God. By the twenty-first century, this comforting certainty had virtually vanished. Why? What caused such a cataclysmic change in attitudes to science and to the world?

Science and Spirituality is the history of the interaction between Western science and faith, and of the sometimes productive and occasionally disastrous ways in which scientists have engaged with religious beliefs and institutions. It details the cultural and intellectual politics that ignited the descriptive 'cause' of science, eventually bringing about its ideological separation from its former ally, the Church.

Journeying from the French Revolution to the present day, and taking in such figures as Francis Bacon, René Descartes, Charles Darwin, Immanuel Kant, Albert Einstein, Mary Shelley and Stephen Hawking, David Knight shows how science evolved from medieval and Renaissance forms of natural theology into the empirical discipline we know today. Focusing on the overthrow of Church and State in revolutionary France, and on the crucial nineteenth-century period when a newly emerging scientific community rendered science culturally accessible, *Science and Spirituality* explores the volatile connection between science and faith and challenges the myth of their being locked in inevitable conflict. The book shows how scientific disenchantment has provided some of our most flexible and powerful metaphors for God, such as the hidden puppet-master and the blind watchmaker, and illustrates the way in which questions of moral and spiritual value continue to intervene in the scientific endeavour.

David Knight is a Professor at Durham University and former President of the British Society for the History of Science.

Science and Spirituality

THE VOLATILE CONNECTION

David Knight

Routledge
Taylor & Francis Group
LONDON AND NEW YORK

For Harry, Cameron, Dominic,
Alexander, Isobel and Bridget

First published 2004
by Routledge
11 New Fetter Lane, London EC4P 4EE

Simultaneously published in the USA and Canada
by Routledge
29 West 35th Street, New York, NY 10001

Routledge is an imprint of the Taylor & Francis Group

© 2004 David Knight

Typeset in Garamond by
M Rules
Printed and bound in Great Britain by
The Cromwell Press, Trowbridge, Wiltshire

All rights reserved. No part of this book may be reprinted or
reproduced or utilised in any form or by any electronic,
mechanical, or other means, now known or hereafter
invented, including photocopying and recording, or in any
information storage or retrieval system, without permission in
writing from the publishers.

British Library Cataloguing in Publication Data
A catalogue record for this book is available from the British Library

Library of Congress Cataloging in Publication Data
Knight, David M.
Science & Spirituality: the volatile connection/David Knight.
p. cm.
Includes bibliographical references and index.
1. Religion and science–History–19th century.
I. Title: Science and Spirituality. II. Title.
BL245.K59 2003
261.5'5'09034–dc21   2003045927

ISBN 0–415–25768–9
ISBN 0–415–25769–7 (pbk.)

Contents

Acknowledgements   vii

1   Something greater than ourselves   1
2   Christian materialism   8
3   Watchmaking   20
4   Wisdom and benevolence   37
5   Genesis and geology   53
6   High-church science   74
7   God working His purpose out?   92
8   Lay sermons   108
9   Knowledge and faith   123
10   Handling chance   137
11   Clergy and clerisy   151
12   Mastering nature   167
13   Meaning and purpose?   183

Notes   197
Index   225

Acknowledgements

This book grew out of a course delivered over the years at Durham University. I am grateful to colleagues and students who have helped me develop it and to the university for giving me research leave in the summer of 2000 to get the book well under way. The Council of the British Society for the History of Science persuaded me to give my Presidential Address in 1995 at the international conference in Imperial College commemorating the death of T.H. Huxley, and I would like to thank them for getting me down to work on him, and the Royal Society and Imperial College for co-sponsoring the conference.

I am also grateful to the Templeton Foundation and to the Center for Theology and the Natural Sciences in Berkeley, California, which awarded the course a prize in 1997, and gave me an award for teaching the following year. As I result, I attended workshops in Chicago and in Toronto, which were most stimulating; and in 1999 was invited to lecture at the Toronto workshop. I would also like to thank the American Anthropological Association – for their invitation to speak at a session on Creationism at their meeting in San Francisco in 1996 – and the Royal Society, which paid my expenses; James Miller and the American Association for the Advancement of Science for inviting me to participate at a session on Science and Religion at their meeting in Anaheim, California in 1999; and David Lindberg and Ron Numbers who invited me to a seminar that spring at Berkeley.

Various parts of this book are the outcome of those and other meetings nearer home. Among many friends who have encouraged me to write, I note especially John Brooke, Geoffrey Cantor and Matthew Eddy; and I am very grateful to all those who came to the conference here in September 2002 on 'Science and Belief' (supported by the Templeton Foundation and the Wellcome Trust) for their support and stimulus at a time when the book was being knocked into its final shape. Finally, my thanks to Gerard M-F Hill for his sympathetic and helpful editing, and to Jackie Butterley for the final task of writing the index.

David M. Knight
Durham, May 2003

CHAPTER 1

Something greater than ourselves

DID GOD MAKE THE WORLD? Does it show signs of intelligent – even benevolent – design? Is scientific study a natural part of religion, part of learning to know God, just as much as reading the Bible? Until about 1800, almost everyone would have answered 'yes' to all these questions. By 1900, many would have answered 'no' to all three, and those who answered 'yes' to the first might well feel awkward about the other two.

Between about the time of the French Revolution (1789) and the outbreak of the First World War (1914), public and academic promotion of 'natural theology' – God's mind visible in His creation – collapsed. At the same time the position of organised religion was weakening all over Europe. We cannot blame that solely, or even chiefly, on scientists; and it was certainly not all the fault of Charles Darwin. The story is more interesting and varied. It is worth knowing: for those who ignore history are doomed to repeat it.

Some thinkers feared that science would simply kill spirituality, but it survived and the two stayed closely together, intertwining and interacting in various ways. Some of the people we shall investigate were the losers; but then, their history is usually more instructive than that of the winners. Britain escaped revolution and invasion, and was religiously diverse, so the story developed differently here, being more drawn out and argued about; and Britain will be our main focus.

Some may think that the tension of 'religion and science' is a recent problem, on which the past casts no light; others may think the problem dates back as far as Darwin. Both analyses lack perspective. The crucial epoch, the long nineteenth century, the age of science, is often perceived (wrongly) as dominated by Darwin – but 'natural theology' had lived uneasily with Christian theology for centuries.

What is spiritual?

Even animals believe that the grass is greener on the other side of the fence. Humans, too, have the urge to discover new places and seek prosperity. Allied to the urge to explore, perhaps, is the urge to ask questions and find answers.

People are also driven by the need to improve themselves, to achieve and outdo. It is human to aspire to something higher, even if we don't quite know what. In the literal sense, it made Hillary and Tenzing climb Everest, and Armstrong and Aldrin fly to the Moon. In the metaphorical sense, it has motivated the greatest athletes, musicians, leaders, millionaires and saints.

The word 'aspiration' has now been devalued to mean buying things you can't afford, but even that retains the sense of 'setting your sights higher'. Many ambitions reach no higher than top of the range, executive, exclusive or personalised; but there is more to a better life than self-indulgence. For many people, living a better life means a more <u>spiritual</u> one. That doesn't always mean <u>religion</u>: the distinction between the two has been around a long time.

Isaiah saw the Lord, high and lifted up; the disciples saw Jesus transfigured on a mountain top; Paul was transported in spirit to the seventh heaven. Such religious experiences are what the word 'spirituality' used to mean: it was applied to Saint Teresa, Saint John of the Cross, and other devout people who saw visions and dreamed dreams. Those who in this way had been to heaven did not doubt the reality of what had happened to them. It confirmed their faith, though often theirs was a less orthodox faith than most people's. Manuals of prayer and devotion, notably the *Spiritual Exercises* of Ignatius Loyola (founder of the Jesuits), encouraged meditation and promoted spiritual awareness.

But intense religious experience was always viewed as dangerous by the powers that be. It might be gnostic, pantheistic or even atheistic, at odds with the beliefs that cool reason, exemplified in catechisms and creeds, could accept. By the nineteenth century, such experience was called 'mysticism'[1] and spirituality was being thought of as something vaguer, 'a sense of something greater and other than myself that adds meaning and value to my life'.[2] This may sound sentimental[3] but most scientists, past and present, would agree with it. As Mary Somerville put it, science is exact calculation and elevated meditation.[4]

If 'something greater' is being given significance, even though it cannot be weighed, measured or empirically verified, that is a straight rejection of the kind of materialism we might expect from scientists. Such rejection need not be connected with a religious tradition, or show itself in formal worship. Most men of science have seen the world as wonderfully and fearfully made, the best of all possible worlds, more complex and sublime than superficial observers would ever suppose. They have often seen it not as a machine, Baruch Spinoza's *natura naturata,* but as dynamic and working, like an organism, *natura naturans.* Here, as in the Neoplatonic tradition, matter seemed alive, active,[5] a sleeping spirit rather than congeries of billiard-ball atoms. Despite that, Christian or pantheistic materialism is not an oxymoron, and it need not be reductive, as we shall see with Joseph Priestley and Humphry Davy.

Science and religion were portrayed by some in the nineteenth century, that watershed with which we shall be particularly concerned, as necessarily in conflict.[6] This image is still popular, but the reality seems much more complex and interesting, better indicated by the word 'engagement'[7] – which covers everything from agreeing to get married to outright warfare. E.H. Gombrich began his famous *Story of Art* (1950) with the declaration that there is no such thing as art: there are only artists. Science, religion and spirituality in the same way are not unchanging entities, but fluid and very different things at different times in sundry places.

Thomas Gisborne's Natural Theology

Thus early in our story, Thomas Gisborne graduated in mathematics at Cambridge in 1780 as 'sixth wrangler'– meaning he had the sixth best mark in the university's most prestigious degree. He became a parson and squire, being appointed in 1823 to a comfortable stall as a prebendary of Durham Cathedral. He was a friend of William Wilberforce and other prominent evangelicals. Responding to William Paley's optimism, in 1818 he published *Natural Theology*,[8] a little book demonstrating what seemed to him the complete harmony between science (where he had tried to keep up to date) and Christianity.

The world for Gisborne is no perfect clock, but a ruin, its order destroyed by the Fall of Adam and Eve and then by Noah's Flood. He saw the evidence all around: the upset geological strata, the existence of disease and pain, volcanoes and earthquakes, the way humans are ill-suited to life in our chilly climate, all these and more indicated that a Creator who had made a Paradise had now in His rightful wrath turned it into a penal settlement (like Botany Bay). Nevertheless, God had foreseen the Fall, and in His mercy had given us the ability to find food and clothing, and to discover more about the world through reason. Eventually we would see that God's characteristic touch was everywhere, internal evidence of His authorship to those who studied the text of the world around them. This natural theology was a kind of overture to the Christian revelation, the Old and New Testament, which allowed us to get back to harmony with God.

Gisborne took the Bible as a whole, a unity, and as a literal statement of what had happened at the beginning of the world and since: he saw no reason to doubt that the Creation happened about six thousand years ago or that the story of the Tower of Babel explained the proliferation of languages. He also indignantly argued that unbelief was sinful (whatever easygoing sceptics might say) in the face of evidence from both nature and the Bible. There are no doubt some who still think like him, but not many among those as well educated as he was: not all his generation were evangelicals, emphasising as he did atonement and salvation by faith alone, but most were. He would not then have seemed at all eccentric.

By 1914 everything looked very different. Even to Gisborne's contemporary William Blake, this patriarchal God was Old Nobodaddy, an old man with a white beard, a forbidding deity who punished with lives of pain descendants over more than twenty generations for the naughtiness of their unsophisticated ancestors in Mesopotamia, and called it justice. Where Blake led, more and more followed, and among the children and grandchildren of Gisborne's generation even the orthodox had to come to terms with the fact that, as a record of events, the Bible was not literally true.

What had come to be read (even in its English translation)[9] as straightforward science and history, like some kind of textbook having divine authority, came to be seen as edited, inconsistent and fallible: inspired, perhaps, but not in some simple or literal way. By the middle of the nineteenth century, there was even the suggestion that Saint John's gospel might be a work of fiction. 'Gospel truth' was thus threatened: just like writers of science, history or biography, the four evangelists might after all be tested against the evidence and found to be erroneous.[10]

The Bible had in earlier centuries been read very differently, not only as narrative but as allegory, where the deep meaning lay beneath the surface: types and prophecies with a bearing on the current state of things, carrying immediate personal or corporate messages. Indeed, in liturgy the Bible is still used and expounded in these ways, different passages being read out to cast light on one another, on the day's theme, and on the congregation's predicaments. But the literal sense had come to be accepted as the one true meaning by the early nineteenth century;[11] and 'biblical criticism' looking at the text with a chilly academic eye gradually spread from Germany to the rest of the world.

With this textual criticism went criticism of the morality of the Bible, both the ruthlessness attributed to God in the Old Testament, and the meek turning of the other cheek in the New. The Bible by 1914 had to be seen as the work of many writers and editors, a challenging collection of stories and poems rather than packaged truth taken down by inspired and inerrant penmen; but those brought up with a strict regard for truth found it hard to accept that 'myth' might be a valuable category for powerful stories that were not simply true or false.[12]

Religion: embarrassing or dismal?

Science went readily with a sense of the sublime, with wonder at the extent and majesty of the world, the intricacies of minute forms of life, and the order and harmony of the whole interdependent universe. But, for the enquiring mind, committed religious belief became more difficult as new ideas trickled down from the French *philosophes* and materialists. Once it reached the French revolutionaries, the perception that religion was a bad rather than a good thing steadily gained ground, until in our day it is probably a majority view in Europe.

Churches came to be seen as dogmatic, inhumane and obstructive: human endeavour did not flourish there, clerical power and status were abused.

Organised religion gave way to the personal kind: believing-but-not-belonging became more common. We are apt to see belief as meaning assent to doctrines, but it may be more helpful to see it as trust: 'We believe in Britain' or 'I believe in her' rather than belief in unverifiable propositions.[13] But in both cases, there is necessarily room for doubt and qualification – we can never be sure of doctrines, and we can never be sure of people. There may come a time when we do not believe in our country, right or wrong, or our faith in a friend may be shaken.

So it has been with miracles. With the rise of science, faith in miracles came to be an embarrassment to many people. God's foresight, wisdom and benevolence were shown in the laws of nature, which were never broken, and sudden interventions seemed unworthy and implausible. Of Mary Somerville it was said[14] 'Above all, in the laws which science unveils step by step, she found ever renewed motives for the love and adoration of their Author and Sustainer'.

Many of those involved in science felt the same way, but this lawgiving God of Nature was a remote figure, more like an efficient administrator. This was not God the Father, Son and Holy Spirit to whom one might pray – who cared about the fall of a sparrow, and much more for our misfortunes. To critics, including the poet–philosopher S.T. Coleridge and the Oxford Tractarians of the 1830s,[15] William Paley's 'evidences' of design seemed to leave little but mere curiosity and wonder, or a dismal religion without spirituality. Later, in Charles Darwin's theory of evolution by natural selection, Nature was red in tooth and claw: could it be said that God was love indeed, and love Creation's final law?[16] The God of science might seem to be the resting God of the seventh day, letting Creation run without interference, not the loving father worthy of personal trust.

Science: a world apart?

We think we know what science is: laboratories, formulae, gobbledygook. At first, though, it just meant knowledge or know-how (from Latin *scientia*), but it came to mean higher knowledge or specialist study in any subject – and in other languages it still does: an Academy of Science may include art and literature. In English, though, it is used mainly for the study of the natural world. 'Knowledge' is perhaps misleading: research may produce anomalous results, the researcher may compensate for variables that cannot be excluded and the theoretician searches for a rule, a law, that will explain everything. Knowledge is really our understanding so far, and it is part observation, part interpretation; in both cases, the results may not be what we expected.

Among intellectuals in the USA, the optimistic belief in the early nineteenth century, that science would always illuminate religion and was a form of spiritual seeking,[17] gave way in mid-century (under the impact especially of evolutionary thinking) to an idea of separate spheres for religion and science, but still with the expectation of long-term reconciliation. These hopes went with the perception that the universe was governed by laws. But then chance, probability and hypothetico-deductive science became ever more prominent in the last decades of the century. As pragmatic acceptance of scientific uncertainty gained ground, agnosticism became an increasingly possible option, though Americans may have been less content with doubt than Europeans.

One way out of doubt in the early twentieth century was fundamentalism, within Christianity a distinctively (though not exclusively) American solution to the problem. In defiance of what they see as godless and materialistic evolutionary science, people have come up with ingenious glosses upon Genesis in order to square it with the fossils and preserve family values which they cherish. 'Social Darwinism' has in fact been used to justify equality of opportunity just as easily, though not as often, as ruthless competition: and it can, as with T.H. Huxley, go with an admiration of the Old Testament prophets and a morality of self-sacrifice.[18] In Britain, Huxley identified Darwinism with populism and opposition to a sclerotic Establishment in church and state, whereas in the USA, after the Civil War, it was identified with snooty Yankees imposing new commercial values upon a defeated South. It was duly hated in what came to be called the Bible Belt.

Traditions and practices are more important in many religious traditions than are beliefs, and the same may be true of science. Ethical questions about science and technology would thus seem to be some of the most profitable ones for outsiders to ask, and it is thus weird that fundamentalists seem more concerned with scientific beliefs than with scientists' and technologists' sometimes alarming practices – perhaps because the Bible does seem to justify exploiting nature. Curiously, because fundamentalism is a modern reaction to feelings of exclusion rather than really the faith of our fathers, it seems readily compatible with physical sciences and technology:[19] but we may perhaps remember Thomas Kuhn's remarks on how dogmatically the sciences are taught.[20]

It is, however, daunting that fundamentalism (Christian, Islamic, Jewish or secular) should go with a technical training as easily as it goes with ignorance: perhaps it is because in such training, especially, *mythos* is downplayed in favour of *logos*, and education is seen as the acquiring of information and skills rather than as a preparation for an examined life, in accordance with reason.

The two cultures

By 1900 science had become a professional and specialised activity, its leaders competing for intellectual authority with sages and clergy. Natural theology, as the scientific sublime, had been used in earlier days by those getting science across to the wider public; by contrast, twentieth-century popularisation was generally more secular, although that century was in fact dominated by belief systems, including nationalism, ideology and religion. Knowledge had become fragmented and specialised.

One result was what C.P. Snow in 1959 called *The Two Cultures*: "Between the two a gulf of mutual incomprehension".[21] As Michael Flanders, introducing the second law of thermodynamics, put it: 'Ordinary people find it difficult to talk to scientists; and scientists simply can't talk to anyone else'. It was now difficult for those involved in theology or spirituality to know who were the real scientific authorities, and conversely for those working in science to know who were the important and representative figures in the churches. Liberal Anglicans between the wars rejoiced in the spiritual-mindedness of Oliver Lodge, James Jeans and Arthur Eddington, and supposed that Charles Raven was an important figure in biological thinking.[22] Looking from the other side, devotees of scientism[23] think that fundamentalist 'creationists' are the really significant religious thinkers.

To know who are the central figures in the science or religion of any generation is hard for outsiders: and it is a real question whether those insiders involved with the trees, or the outsiders looking for the wood (the big picture), are the more important – or better placed to judge. We tend to under-rate the popularisers of previous generations, even though we depend on those of our own. History of science tends to concentrate on the pioneers, discoverers and inventors; so we expect to find those great figures even more important than they were. Then as now, it was accessible rather than very demanding books and papers that were read.

Our story begins in the late eighteenth century, with Joseph Priestley and the epoch of the French Revolution. It was easier then to know who was who and what was what, because there was only one high culture and the intellectual world was small: most intellectuals knew each other, or at least knew of each other. And we shall find Priestley's especial and idiosyncratic synthesis of science and religious belief well worth investigating, overtaken though it was by political events.

CHAPTER 2

Christian materialism

ON 14 JULY 1789 THE BASTILLE fell. It had stood as a terrifying monument to absolute power, and its storming and demolition marked

> A time when Europe was rejoiced,
> France standing on the top of golden hours,
> And human nature seeming born again.[1]

The word 'revolution', as used in Britain in 1688 and in America in 1776, implied a turning back to good old days of Merry England, before liberty was infringed by tyranny. As such, the events in France were welcomed abroad; but things soon turned serious.

The French state and church were so entangled that the fall of one brought down the other. The overthrow of constitutional monarchy turned into republican violence, with the execution of the king and queen, and the Reign of Terror. Beset by foreign armies, the republic fought back and began to export liberty by force, to the Netherlands, Italy and Germany. The conquered paid for further conquests. Britain joined in and in 1794 a world war broke out, with fighting in India and the USA as well as in Europe: it continued, with two brief intermissions, until the battle of Waterloo in 1815.

These events were shocking in Britain, where enlightenment had taken 'the form of a century's struggle to understand the harmony of religion and science';[2] and were blamed on the *philosophes* responsible for the French Enlightenment: Denis Diderot, Jean d'Alembert, Jean-Jacques Rousseau and Voltaire. Their writings – scornful of the present, idealistic about the remote past or the future – were seen as an ideology of revolution; and Claude-Adrien Helvetius and Paul (Baron) Holbach were especially marked out as atheists and materialists. Atheism in the eighteenth century, and in Britain through most of the nineteenth, was usually seen as irreligion, not a speculative belief that there was no God. It meant indifference to the Ten Commandments and the Last Judgment. Few therefore admitted to atheism, though many were accused of it.

Helvetius and Holbach were different, passing beyond scoffing and hedonism to serious atheism: and the consequences were just what learned divines had predicted: that the collapse of organised religion would lead to the downfall of morality and social order, to the Terror and military dictatorship. A whole generation grew up for whom war was normal. Men of science in France devised methods for melting down church bells into guns. A flood of emigrés poured across the Channel to Britain, including not only aristocrats but also priests who would not swear oaths of allegiance to the new government rather than to the Pope. The Roman Catholic seminary set up at Douai in the sixteenth century, to train priests for England, now fled France and set up at Ushaw near Durham. Although Gallic elegance was still admired, and her secular modernity and perhaps meritocracy envied by some, politically unstable and militaristic France became for most Britons the enemy, distrusted and feared right through the nineteenth century.

The French Revolution indeed marked a discontinuity in modern history: things never looked the same after it. It was not a return to the Good Old Days. The French empire was liberated from kingcraft and priestcraft; and while there was some return to monarchy, the status of religion was never again the same. Napoleon's armies set Europe on the way to becoming a secular continent.

Priestley and materialism

Joseph Priestley was a friend of Benjamin Franklin and Thomas Jefferson, a chemist, minister of religion and left-wing politician. He sympathised with the Revolution and was even elected to the French National Assembly, though he did not take up the seat. Since 1689 there had been religious toleration in Britain, at least for Protestants: those who disliked the established church (which was Episcopalian in England, Presbyterian in Scotland) had a wide choice of sects, with or without ordained clergy. Dissenters had ministers chosen and paid by their congregations. In France, to abandon the Roman Catholic Church (unless one went underground as a Protestant) was to abandon the Christian religion.

Priestley grew up as a studious boy in a Dissenting, Calvinist family, and the intention was that he should enter the ministry. The universities of Oxford and Cambridge, as Church of England corporations, were closed to conscientious Dissenters, so men like Priestley attended Dissenting academies – which trained ministers, and also those going into business. Many Dissenters stoutly believed in education, and in getting on, and became prominent in industry and commerce: but Dissenters were still second-class citizens, with civic positions closed to them. Some – like Priestley – bitterly resented being thus stigmatised, and looked forward to an England in which no church was privileged and religion was a private matter. At the same time, Priestley found himself unable to accept the strict Calvinism of his youth, and moved in a liberal and unorthodox direction.[3]

What then is materialism? 'If you've only got one life, why not live it as a blonde?' ran an advertisement for hydrogen peroxide half a century ago, and that summarises materialism, as most people see it. Why not eat, drink and be merry, if tomorrow we die and that's that? Duty, sin, judgment, morality, perhaps even justice lose their force if we are just a mass of particles that fall apart at death. If there is nothing beyond what we can sense, the gratification of our desires might appear the only sensible course. The great and the good naturally denounced materialism. Priestley saw it differently.

Reading and reflection convinced Priestley that (as Newton had secretly believed) Jesus was not God, that the doctrine of the Trinity was a 'corruption' of Christianity from Platonic philosophy, and that his 'Unitarian' belief was the true faith of Jesus and his disciples. His studies indicated to Priestley that to early Christians, and indeed to himself, Jesus had not seemed uniquely the Son of God; the Son of Man (as he had called himself) was a man. Priestley believed the separation of matter and spirit to be contrary to genuine Christianity. Matter had associated with it, after all, powers of attraction and repulsion; and there was no reason to suppose that matter, suitably organised, could not think. The notion of utterly distinct bodies and souls was anathema to him – another Platonic corruption, which had crept in to Christianity. We were not eternal souls, imprisoned in material bodies, souls which at death would be liberated to go to heaven, or hell: we were composed of matter that at our death would rot. By a miracle, God would at the Day of Judgment raise us up bodily.

The true Christian doctrine was thus for Priestley not the immortality of the soul, but the resurrection of the body. This is indeed just what we see in some medieval drama, such as the York mystery play of Doomsday, or in Renaissance pictures, including the east wall of the Sistine Chapel: as the angels blow their trumpets the graves are opening, and the awaking dead coming forth to face Jesus as judge. Contrariwise, in the York 'Harrowing of Hell' and in Dante's *Inferno* the immortal dead await rescue or have already been condemned. The Apostles' Creed speaks of the resurrection of the body; the Nicene of the resurrection of the dead. It is a curiosity of history that popular belief went over so firmly to the doctrine of the immortal soul, with death as family reunion, by the early nineteenth century.[4]

For Priestley, therefore, Christianity was materialistic; and this provided him with a basis for uniting his science with his faith in what seemed to him a thoroughly satisfying compound. Because he had a stutter, he could never make a great preacher; but he found a congregation, and honed his writing. Finding himself living next to a brewery in Leeds, and intent upon learning about God through His works, he duly investigated the air that frothed up in the vats. It turned out to be 'fixed air' (our carbon dioxide), also found combined in limestone; and this launched him on a life of chemical discoveries, which included the invention of soda water (thus founding the soft-drinks industry) and the isolation of oxygen.

He worked, following Franklin, on electricity; and invented the time chart in an important contribution to history. His book on electricity declared that:

> Hitherto philosophy has been chiefly conversant about the more sensible properties of bodies; electricity, together with chymistry, and the doctrines of light and colours, seems to be giving us an inlet into their internal structure, on which all their sensible properties depend. By pursuing this new light, therefore, the bounds of natural science may be extended, beyond what we can now form an idea of. New worlds may open to our view, and the glory of the great Sir Isaac Newton himself, and all his contemporaries, be eclipsed, by a new set of philosophers, in quite a new field of speculation.[5]

His writings brought him into controversy with orthodox clergy, and with the conservative thinker Edmund Burke – one of the first to see which way events in France were moving.

Atoms come from Croatia

In the later *Disquisitions Relating to Matter and Spirit*[6] (the Preface is a brief intellectual autobiography)[7] Priestley set out a theory of matter that he had adapted from that of a Croatian Jesuit, Roger Boscovich. He could not accept that matter was composed of inert atoms, Newton's 'solid, massy, hard, impenetrable, moveable Particles' formed by God in the Beginning[8] and surrounded by void. How could such particles act upon each other, as they do under gravity? Even more, how could they stick together to form solid bodies rather than mere heaps?

Newton himself did not know the cause of gravity, and sometimes attributed it to God's direct and continuing action. Some of his contemporaries saw immaterial substance, akin to spirit, as the glue. Newton's great rival and critic, Gottfried Wilhelm Leibniz, mocked as a 'perpetual miracle' the idea that matter might act at a distance across void space, as the Sun attracts the planets and comets in Newtonian physics; and the retort of Newtonians that miracles were of their nature occasional rather than perpetual did not really answer the problem.

Boscovich had suggested that atoms were mere points, centres of alternately attractive and repulsive force extending ever more weakly to infinity. These atomic particles – like the minute dots which make up the letters of a printed text – were held together in very stable arrangements like our chemical elements, at their nearest attractive distances. In chemical combinations the distances were greater; while gravity operated at long range. At the centre the repulsive force rose to infinity, so no two atoms could occupy the same space. What seemed to us mass was really weight, the effect of a force: and force was all that we ever really met, though it might seem solid and (as Humpty Dumpty found) impenetrable.

Boscovich believed that his theory allowed him to demystify Newtonian attraction, while remaining completely orthodox in his religious beliefs. He visited England, and conversed in Latin with Samuel Johnson; and his ideas gained a certain currency. Priestley came across them when he was employed by Lord Shelburne, a prominent if unpopular politician – briefly Prime Minister – a supporter of Dissenters and American independence. The recently-widowed Shelburne made him tutor to his sons, librarian at his mansion at Bowood, and general intellectual companion;[9] giving him leisure to write and to carry on research. Priestley gained the opportunity to visit pre-Revolutionary Paris when Shelburne went there on a diplomatic mission: he met the great chemist Antoine Lavoisier and told him about his work on gases, putting him on track for the oxygen theory of burning and what Lavoisier called his revolution in chemistry – his book was about to be published in 1789. Chemistry was the most exciting science of the day, with a new language and a new way of looking at combustion, acidity, and the gaseous, solid and liquid states.

Priestley met at a dinner many of the leading men of science in France, for Paris was from the 1780s to about 1820 the world's centre of excellence right across the sciences. He was told that, though many of them held ecclesiastical offices in the Roman Catholic Church, of course they were not believers; to which Priestley indignantly retorted that he was a Christian. He was careful in his book to say that God was not material, though everything else is. But for Shelburne the *Disquisitions* was a book too far. Priestley's use of the point–atom theory to get rid of immaterial substance, and hence of spirit, began to make him an embarrassment to Shelburne; and Boscovich was publicly furious that his ideas had been used to support infidelity.

Birmingham and exile

The favourites or sidekicks of the great must expect to be sacrificed when things are not going well: Shelburne had married again, and his sons were growing up; and Priestley had to find himself a job. He was called to a congregation in Birmingham, where James Watt and Matthew Boulton were making steam engines. There he became a leading light in the Lunar Society,[10] a group which met at the full moon (so that they didn't have to go home in the dark) and included Boulton and Watt, the physician Erasmus Darwin and the potter Josiah Wedgwood, who made apparatus for Priestley and became a Unitarian.

Indeed, in the Darwin and Wedgwood families Unitarianism was called 'a featherbed to catch a falling Christian'.[11] Unitarianism[12] as an ultra-liberal and very intellectual form of Dissent won many converts among the well-educated; and Unitarians were prominent in the Literary and Philosophical Societies, Academies and Athenaeums which formed in provincial towns and cities, notably in Manchester and Newcastle upon Tyne. Priestley continued to write religious and philosophical controversy, coining the phrase 'the greatest happiness of the greatest number' as part of his utilitarian moral philosophy, later taken up by Jeremy Bentham and his followers. He was a well-known figure, often caricatured.

Then came the French Revolution. Priestley, like most of his fellow-Lunars, was a fervent supporter, though he later came to deplore the Terror – in which Lavoisier lost his life on the guillotine as an indecently-rich tax-collector. Priestley had hoped for a reformed and secular government in Britain, more like that achieved in America or – he believed – being set up in France. On 14 July 1791 he was (wrongly) supposed to be involved with a celebration of the anniversary of the fall of the Bastille; and a mob, shouting 'Church and king' went to his Birmingham house and laboratory, and sacked it – luckily he had been warned in time, and fled. Other radical dissenters trembled. Perhaps as a consequence, in the nineteenth century it was Manchester rather than Birmingham that made the running as an intellectual centre, especially among Dissenters – the catchphrase was 'What Manchester thinks today, London will think tomorrow'.

Priestley himself went to London, where he taught at the Dissenting Academy in Hackney – at that time a hotbed not only of intellectual dissent, but also of a group of Anglican high-churchmen called the Hackney Phalanx. Eventually Priestley extracted from a reluctant government compensation for his losses; but he was not made to feel welcome in the Royal Society, of which he was a distinguished Fellow. The French had a salaried Academy of Sciences (of which Priestley had the distinction of being a Foreign Member), but the Royal Society was more like a club, few of whose members actually undertook scientific research. As attitudes to France hardened, especially among the governing class (of which the Royal Society was composed), nobody wanted to know Priestley.

In the summer of 1794 he left for Pennsylvania; where he lived by the Susquehanna River for the last ten years of his life, under some suspicion even there at first until Jefferson became President. He continued to resist Lavoisier's oxygen theory, believing that it did not measure up to the proper Newtonian standards of proof: the older phlogiston theory (anything that burnt contained this rather mysterious component) had been his guide, along with what he saw as a strict empiricism. He had always believed that science was not recondite or too difficult for most people, and credited himself only with a prepared mind – unlike Lavoisier, who was very much the elitist and the theoretician.[13] It was a sad thing that the two greatest chemists of the age should be condemned, into exile and death, by the societies with which they identified themselves.

A cool religion for the comfortably off

Priestley's rational Dissent and his belief in science might seem unsuited to a 'godfather' of the first generation of Romantics. But Thomas Beddoes, his chemical disciple, set up with Lunar support in Bristol[14] a clinic where the 'factitious airs' isolated by Priestley could be administered to the sick, especially sufferers from tuberculosis. High hopes were entertained for oxygen, and the properties of laughing gas were discovered there by the young Humphry Davy, Beddoes' assistant.[15] There is an encouraging letter of 1801[16] from

CHRISTIAN MATERIALISM

Priestley in America to Davy, upon whom he clearly felt that his mantle had fallen. Those dynamical sciences that Priestley had picked out, chemistry, electricity and optics, remained exciting for romantic thinkers – though not Newtonian mechanics.[17]

Beddoes himself had recently been the first reviewer of Immanuel Kant's *Critique of Pure Reason* for an English journal, the *Monthly Review*, and perceived that there was exciting thinking and writing going on in Germany though he was unsympathetic to it in his book on Demonstrative Evidence.[18] He encouraged William Wordsworth and Samuel Taylor Coleridge to visit Germany. When Coleridge returned, and he and Robert Southey married sisters, they resolved to form a commune to be called a Pantisocracy on the banks of the Susquehanna River, chosen partly no doubt for its beautiful name, but also because of its proximity to Priestley. Coleridge himself intended to enter the Unitarian ministry, *faute de mieux*; but he was saved from clerical life by an annuity from the Wedgwoods. Later he attended Davy's lectures in London to improve his stock of metaphors.

Coleridge and Wordsworth when writing the *Lyrical Ballads* (1798) had been suspected of political radicalism, and put under surveillance; while Beddoes' French sympathies seemed proved when he was seen collecting frogs for electrical experiments, and was naturally thought to be harbouring and feeding a spy. But even they moved to the right, disillusioned. Political reaction set in during the 1790s, and demanding political change was unpatriotic. British politics remained frozen for more than a quarter of a century. Now that radicalism seemed discredited, Priestley's liberal vision of the union of science with true religion seemed discredited too. There were still many Unitarians, but it was the Methodists who converted the masses.

Unitarianism seemed cool and rational, a religion for the comfortably off and well-educated, while Methodism was a matter of the heart – with John Wesley's enthusiastic sermons promising salvation to all who claimed it, and Charles Wesley's hymns transforming church services generally. Unitarianism, Priestley thought, went well with science. It was an enlightened faith, that excluded miracles and mysteries, focused upon the wisdom of God the Creator and First Cause, and urged benevolence and intellectual curiosity. It interpreted the Bible liberally: where a literal reading conflicted with up-to-date empirical findings or moral judgments, it was to be taken in some other sense. As the Arian and then the Socinian heresy, Unitarianism was familiar to orthodox clergy, who vigorously denounced it. Unitarians, however, claimed that theirs was the true form of Christianity, and were offended not be recognised as Christians.

To outsiders it looked nevertheless as if Unitarians had come close to sacrificing true religion, believing only that the world had a First Cause, with a morality based upon utility and judgements of the probable consequences of actions, rather than whether they were right in themselves or associated with God's purposes. Unitarianism, as that 'featherbed' nickname implied, looked

14

like a comfortable intellectual position, and a church in which one might meet congenial people; but which was only a short distance from unbelief, and where spirituality was downplayed. Because established churches did not examine the beliefs of their lay members too closely, it is possible that some in the Church of England shared Unitarian ideas. But Priestley's true Christianity with materialism failed to persuade in the hostile climate of the 1790s. If science was to be taken seriously and admired by respectable citizens, it would have to be very different from what Helvetius and Holbach had preached. Materialism remained a horror word.

There is a God – but no point in praying

There were few atheists, but ever since the seventeenth century there had been Deists, for whom God was an impersonal First Cause. Deists pursued the sciences as a religious duty, as demonstrating how wonderful and admirable was the design behind the world; but there was no point in praying to their God, who had designed and launched the best of all possible worlds and could not be expected to tinker with it for the benefit of intercessors. Indeed, any such tinkering would in the broad and long view be a misfortune, because it would make the world less than best. There was little consolation here for the widow and the orphan; and no place for Revelation. Nature is a better guide to God than is the Bible.

Equally, there is no special reason to go to church and worship together: lectures on astronomy, medicine or chemistry might be the best form of service. Deism was characteristic of many intellectuals in what we can loosely call the Enlightenment (for it seems that there were several enlightenments in different countries, with rather different characteristics); but Priestley's Unitarianism is by comparison warm, human and satisfying. By 1800 it seems likely that there were few Deists left in Britain – the failure of the French Revolution to deliver peace and prosperity was a blow also to Enlightenment optimism.

Some have seen the collapse of Deism as a sign of a tendency towards atheism among the English Romantics,[19] a native trend unappreciated by scholars dazzled by the German input. But, while infidelity was indeed an accusation made against many intellectuals, it could easily be applied to Deists or Unitarians too – but they didn't see themselves as atheists. Just as we use the word 'infidelity' in connection with marriage rather than baptismal vows, so then it was expected that giving up the faith would lead to all sorts of other infidelities. Atheism was expected to mean amorality, a charge which all but a very few – including the Sceptics – rejected with asperity.

Nevertheless, there was a good deal of scepticism, of the sort we associate with David Hume and Edward Gibbon.[20] Its basis was that nothing certain could be said about what was unobservable and untestable. Denial of God's existence was as ill-advised as assertion of it – which may sound rather like atheism, but really went with an anti-dogmatic stance on other questions, for example in science and history. A suspicion of 'systems' and a devotion to empiricism marked the sceptic; and the Royal Society's motto, *Nullius in verba* – 'Take nobody's word for anything' – is a good expression of the sceptical attitude. Sceptics might also feel that organised religion was a good thing, because it damped down and canalised a potentially explosive phenomenon. This might sound like 'religion as the opiate of the people'; but it can be a serious position, because fanatics are dangerous people. The history of fundamentalism in the USA might make us feel likewise that an established church is no bad thing.

Sir Joseph Banks

After the Birmingham riot, Priestley hoped for support from Sir Joseph Banks, since 1778 President of the Royal Society, which he was to rule until his death in 1820.[21] Davy, his successor, was born a few days after Banks' election. Banks became famous because, instead of going on the Grand Tour of Europe, he had travelled with Captain James Cook aboard HMS *Endeavour* on her voyage to Tahiti, to follow the transit of Venus across the Sun, and then to look for the unknown southern continent.[22] He was a keen botanist, and took with him Daniel Solander, a pupil of Linnaeus, and two artists (one for landscapes, the other for natural history).

Banks was also very interested in what we would call anthropology, quick to observe and even to enter into the customs of other societies. Indeed his sex-life on Tahiti was a subject of jest for years after his return to England. He gave the Tahitians Latin nicknames because of their supposed resemblances to classical heroes. He was appalled by the primitive lifestyle of the Fuegians near Cape Horn; and taken aback by the utter nakedness of the Australian aborigines. It was Banks who botanised in Botany Bay and who urged the government that, since America no longer wanted British convicts, that transportation to Australia would be a good plan. He later recommended that Merino sheep, a flock of which he had reared for King George III at Kew, might do well in New South Wales. He thus laid the foundations both for settlement and for the Australian economy.

Banks hoped to go with Cook to the Pacific again, this time with musicians and other staff. When the enlarged cabins to accommodate them made the ship 'crank' (unseaworthy), Cook ordered them removed and Banks walked out in a huff. After a visit to Iceland, he settled down in England and decided to devote his life to scientific administration rather than, as might have been expected of one in his station, to politics or a country estate.

The best of all possible societies?

Banks was an important public figure, bridging any gap between sciences and arts, the chief figure in what can be called the English Enlightenment.[23] His

CHRISTIAN MATERIALISM

scientific interests were essentially in natural history, but as President of the Royal Society he seems to have been fair – though mathematicians suspected him of prejudice. Like Francis Bacon in the early seventeenth century, he had a vision of science as carefully established useful knowledge. Banks was not very interested in 'abstruse discoveries beyond the understanding of unlearned people'.[24]

Bacon too had preached an accessible and useful vision of science. He believed that the two books, the Bible and Nature, could never be in conflict if read aright; and that the secret in science was to beware the 'idols' of preconceptions of all kinds, to observe and experiment carefully and open-mindedly, and to generalise cautiously.[25] If this were done, in place of the dogmatic and metaphysical reasoning in which Bacon believed his predecessors had indulged, then the knowledge achieved would help mankind to undo the worst effects of the Fall of Adam and Eve, and the curse of work and pain.

As president, Banks was the chief spokesman for science, to whom secretaries of state and prime ministers turned: there were no others of comparable eminence in science.[26] He was made a privy councillor, giving him privileged access to government; he was Lord-Lieutenant of his county for a time; and his landed wealth made him a weighty person. He saw off a revolt in the Royal Society, and for the last thirty years of his life was unchallenged in his gouty despotism: his achievement was to guide British science through the years of war, when it appeared (as he put it) that if men of science did not hang together, they would hang separately.

He placed science emphatically on the side of good order, of the established social set-up, though it could function also as a vehicle for mobility: Davy, son of an unemployed wood-carver in Cornwall, became Banks' protégé, his successor and a baronet (Newton had been only a knight). Banks functioned, as his society did, through patronage: it mattered more whom you knew than what you knew. But he sought out and promoted talent, and this formidable man was surprisingly tolerant (through his feeling of *noblesse oblige*) of the importunities and insolences of his social inferiors.

Lavoisier, who knew he was a most exceptional person, presided over the Academy of Sciences, and commissioned the latest and most splendid apparatus. Priestley was, by contrast, a humble man, but Baconian science seemed to him democratic and levelling. Experiments should be accurately and rapidly reported so that everybody could join in.[27] Apparatus need not be prohibitively expensive; truth was, for this good Protestant, available for all to find – and it would set them free. Then the conclusions of science, and the industry using science, might well upset tradition and established order, for they would be on the side of Rational Dissent rather than the establishment. The *ancien régime* in France had collapsed because it was incompatible with modern thinking and Priestley hoped that the same might happen in Britain.

Banks shared the Baconian vision but drew different conclusions. True science was, unlike that of the French *philosophes*, unthreatening and socially cohesive. It was also readily compatible with religion. Unlike Priestley, Banks does not seem to have been a religious man. Like many of his eminent contemporaries, he was sceptical, but he believed strongly in the social order and dreaded the consequences of anarchy. He would have deplored religious enthusiasm; naturally he was not likely to go out of his way, or to rock the scientific boat, to help Priestley, whose ideas alarmed the powers-that-be and the conservative.

Divisions in science

People often speak of 'religion and science' as opposites, especially since the publication of John William Draper's *History of the Conflict between Religion and Science* of 1874.[28] Skirmishes over the Book of Genesis (setting scientific theory against unsophisticated Bible reading) may well be sterile, but the applications of science should certainly be kept under moral and political constant surveillance. In doing this, it may be more helpful to think of 'spirituality and science', categories familiar to those who have no part in organised religion. We are not talking here just about 'new age' spirituality: we shall be considering any search for, and interpretation of, experiences which bring meaning to life and which cannot readily – for those involved – be reduced to physiology or psychology.

In the period around 1800, with the rise of Methodism in the English-speaking world, and the shock of revolution and conquest on the continent of Europe, experiences of conversion and the perception of religion as emotional combined with Romanticism to emphasise spirituality. Scepticism, Deism and even Unitarianism all lacked it: they were systems of belief for the cool and hard-headed.

Priestley hoped that everyone would be attracted to science (and then to Radical Dissent); Banks was less sanguine. But to maintain the momentum of science, to challenge the supremacy of Paris in the scientific league, to win the war, and to show that science and everything else would flourish best under the splendid and free British constitution, he needed to safeguard the Royal Society. With bishops, generals, admirals and aristocrats among its members, as well as clergy, doctors and lawyers, there was little reason to fear that it would become a revolutionary body.

But the government was suspicious of more plebeian groups and societies, which might well be radical; and during the 1790s legislation bore down upon them. Banks opposed them, but also specialised societies – for geologists, 'animal chemists' and astronomers. The geologists nevertheless went ahead; but the others became instead subsets of the Royal Society.

It was not only intellectual imperialism that made Banks oppose any fragmentation of science, any development not beneath the aegis of his Royal Society. Only if all men of science were fully included within the Society would it have the weight and the breadth of view, Banks believed, to sustain its meetings and publications, and its authority. He perceived that for many people, natural philosophers were comic figures – absent-minded professors engaged in recondite researches, harmless but probably useless – but they could also be seen as dangerous fanatics. Like Priestley, they might pose a real threat and provoke a reaction threatening Banks and his friends. Science must be the handmaid of established religion.

Volcanoes and natural theology

Burke had famously distinguished the sublime and the beautiful, and his contemporaries recognised that they delighted in the alarming as well as the peaceful. Unlike Elijah, they found God in earthquake, wind and fire, as well as in the still small voice of conscience[29] or the peaceful meadows that had delighted earlier sensibilities, puzzled that God could have designed rocky and uninhabitably horrid mountains. Volcanoes had excited not only Cook and his officers, and their readers, but also Banks' friend Sir William Hamilton who as British ambassador in Naples studied Vesuvius over a number of years.[30] His wife was Nelson's mistress and when the Royal Navy came into harbour, she no doubt despatched her scientific-minded husband off to the mountain with a picnic to do some further studies.

But Hamilton made the study of geology prominent and helped to stimulate discussion about the age of the Earth, and the processes that had modified its surface. The great historians of the Enlightenment – Voltaire, Hume and Gibbon – had aroused their readers to appreciate change over time. The best of all possible worlds, where any change could only be for the worse, was mocked by Voltaire in *Candide*; and the nineteenth century became not only the Age of Science[31] but also the Age of History – which weighed down the living like an Alp.

Priestley's wide interests had taken in respiration and photosynthesis, the cycles involving his fixed and dephlogisticated airs (carbon dioxide and oxygen), but chemistry was his forte – a science whose practitioners improve upon the world (by making new compounds) rather than just admire it. For the Christian chemist, God has given us our wits not just to admire and understand, but to be active and to perfect (as the alchemist had hoped to do) the Creation, by isolating and administering laughing gas or by inventing plastics. The chemist thus creates his own subjects of study.

This made chemistry hard to fit into the programme of natural theology,[32] where contemplating the heavens or the eye of a fly led to admiration of the Design and Contrivance that lay behind them. When Priestley had to flee the country, and his synthesis of science and religion became highly suspect, this older programme of Natural Theology became prominent again, notably in the writings of William Paley; and it is to him that we now turn.

CHAPTER 3

Watchmaking

THERE IS NOTHING NEW IN the idea that the world is so complex and yet coherent that it cannot be accidental. Nor is the idea just a Christian one: Plato, in his last dialogue *The Laws*, inferred the existence of a Designer from the orderly behaviour of the planets and of nature generally. It was design, not chance. His pupil Aristotle, a great biologist as well as a philosopher, took the same line. Animals and plants reproduce their kind, and such predictability and adaptation to life are not just happy accidents – serendipity does not happen over and over again. The serious thinker had to accept that we were the outcome of thought – that we had a maker – and likewise the world around us. This Design argument might be reinforced by religious experience, as in the book of Job, but the argument was – indeed, is – based in logic.

Plato and Aristotle had a target for their arguments. Leucippus and Democritus[1] had supposed that everything was composed of atoms, which hooked and unhooked themselves in colliding: forming worlds, people, animals and plants, which endured for a time and then came to bits. We would thus inevitably fall apart in due course, and that would be the end of us. In a sense we are the product of chance, and so is everything around us; but everything happens according to inexorable law. Even the things that give subjective interest to life are simply arrangements of atoms in the void: 'By convention are sweet and bitter, hot and cold, by convention is colour; in truth are atoms and the void'. These ideas of the structure of matter were taken up by Epicurus and Lucretius to form a materialistic world-view, where the gods watched the world from the sidelines, but were powerless to intervene in the ceaseless dance of the particles, coupling and uncoupling.

But to most people it wasn't enough. Even in the Renaissance, the fascinating style of Lucretius' great poem *De Rerum Natura* and the enigmatic fragments of Democritus could not compensate for their improbability. People just didn't believe that the world was devoid of design and purpose: the world seemed chancy, but not that chancy. At the same time, this powerful picture of a world made entirely of tiny particles – taken up with reservations in the seventeenth century by Galileo, Newton and Locke – led to a separation of the primary qualities (such as mass and size) really inherent in matter from the secondary qualities (colour, taste, smell and so on) that we impose through our senses.

Boyle and Bentley: What makes the universe tick?

The seventeenth-century chemist Robert Boyle was one of the many in his time who sought to reconcile atomic theory with Design. On his Grand Tour he made sure to include Strasbourg, with its wonderful clock. Cathedrals were proud of their clocks, often the only mechanical timekeeper in their city. The Strasbourg one not only showed the time, but also had figures that struck the bells, apostles who processed at noon, and faces displaying the phases of the Moon. After some rebuilds, it still strikes every day at 12 noon, local solar time, and remains a tourist attraction. Today we may see the splendid but rather creaky action of the clock as quaint, but Boyle's generation felt real wonder, because it seemed like a world in miniature. If a human craftsman could design and build such a device, then it was a short step to an all-knowing Craftsman who could construct a world.

Clockwork could also, perhaps less happily, provide the analogy for the smooth running of a state, perhaps one of the modern nation-states in Europe. It went especially well with the absolutism of Louis XIV of France (imitated with dire results by Charles I and James II in England).[2] Anyway, the object of science was to discover, to take the lid off and reveal the mechanisms that made things tick – taking the back off the world-clock, as it were, and peering inside.

Boyle wrote enthusiastically about the potential of mechanical explanation. He firmly believed that sober, inductive science would provide more and more evidence for the existence, wisdom and goodness of God; and at his death left money to fund lectures on this theme. A bright young Cambridge man, Richard Bentley, delivered the first lectures in 1692/3.[3] Bentley had heard about Newton's *Principia*, published in 1687, which set out the theory of universal gravitation and the motion of the planets, on a basis of laws of mechanics that applied to everything.

Newton's masterpiece explained the elliptical orbits that Johannes Kepler had found planets to follow, and so did away with the cumbersome system of epicycles, wheels within wheels, which had been used to account for and predict planetary positions ever since Ptolemy in the first century AD. Now their arrangement in the solar system looked clear and consistent, worthy of a Divine Designer. This was what Bentley realised: the universe was orderly, and not the chaos that would result from chance collisions.

A world of order

Atheism could be refuted. But it was not easy to understand Newton's book; and after a friend recommended a daunting course of mathematical study,

Bentley instead boldly wrote to Newton, asking him about its implications. (Their resulting correspondence was much later published, in 1756, provoking the remark by Samuel Johnson that it showed the mind of Newton gaining ground upon the dark.)

Bentley drew Newton out, making him declare his position: Newton was happy, as he noted in subsequent editions of the *Principia*, to have his new physics used to demonstrate God's existence and wisdom; and he could not accept that powers such as gravity could be inherent in 'inanimate brute matter'. He believed that the planets must have been placed in their orbits, all going the same way round the Sun and in a plane; and that the Solar System could not have resulted from chance processes in particles of matter that had started out evenly distributed. Armed with ideas from Newton himself, Bentley went into his lectures with gusto, emphasising the vast improbability involved in atheism. The lectures were a great success, Bentley affirming glowingly that atheists now hid their heads, and proclaimed themselves Deists instead. Whether this was in practice any better, he did not ponder. A glittering academic career as Master of Trinity College, Cambridge opened before him.

Bentley's astro-theology was followed by a whole series of similar ones; but the same principle – that an ordered universe was not the result of chance – could be applied to the world around us. Even before Bentley's first set of Boyle lectures, another Cambridge man, John Ray,[4] published *The Wisdom of God Manifested in the Works of the Creation* in 1691. He was a very distinguished natural historian, especially for his natural system of classifying organisms, in the tradition of Aristotle. He had been (as Newton was to be) a Fellow of Trinity College, Cambridge; but scruples about taking the oaths required in 1662 after the Restoration of King Charles II made him resign his Fellowship, and give up his position in the established church. Newton more prudently kept his doubts about the church and its doctrine to himself.

God revealed in the eye of a fly

Like Newton, Ray was connected with the group of thinkers called the Cambridge Platonists, who applied the idealistic and mathematical way of their Greek hero to modern problems. Unlike Boyle,[5] who had been reluctant to allow anything to come between God and the world (or between God and his worshipper – he was a keen Protestant), Ray believed that the Creation had been achieved – and was maintained – by a Plastick Power or Demiurge.[6] This, God's Vicegerent or Mother Nature, might make mistakes sometimes, as when unviable monsters are born or disasters occur: so God Himself need not be blamed for everything that goes wrong.

Ray's work on animals and plants convinced him that there was nothing left to chance in them. As Aristotle had said, Nature did nothing in vain, to no purpose. The invention of the telescope had revealed fresh wonders in the starry heavens and those looking through the microscope could now marvel at minute anatomical structures such as the eye of a fly or the legs of a flea. Just as Galileo, Newton and Bentley had made sense of the skies, so too a new territory of physico-theology opened up before the astonished eyes of readers at the end of the seventeenth century, whose ancestors had simply not realised how amazing the world was.

Science was disclosing the wisdom of God. And indeed, about 1700 this was one of its major uses: Francis Bacon's hope that knowledge would be power was not realised until the industrial revolution. Bacon observed that two books, the Bible and Nature, told us about God and this idea gripped contemporaries of Newton, Bentley and Ray. And it was not just an English phenomenon: in very different traditions, Cotton Mather in Massachusetts[7] and Abbé Pluche[8] in France, and Leonhard Euler in his amazingly wide-ranging *Letters to a German Princess*[9] also wrote influential works of natural theology.

This revived emphasis upon bringing natural theology to the public was a major enterprise throughout the eighteenth century. Listening to lectures on science became fashionable,[10] but speakers on lecture tours also found that audiences responded to an emphasis on the goodness and wisdom of God. William Derham was one of the clearest writers in the genre: in 1711/12 he delivered the Boyle Lectures, which were published as *Physico-Theology* and went on selling in new editions through the century. His *Astro-Theology* was equally successful, making him at once an apologist and a populariser of science, for the books contain a great deal of natural history and astronomy painlessly presented. Modern readers may smile, but his reflections upon vipers and other poisonous snakes had reason behind them:

> There would be no injustice for God to make a set of such noxious creatures, as rods and scourges to execute the divine chastisements upon ungrateful and sinful men. And I am apt to think that the nations which know not God, are the most annoyed with those noxious reptiles, and other pernicious creatures. As to the animals, their poison is, no doubt, of some great and special use to themselves.[11]

The Irish, freed of snakes by Saint Patrick, could feel complacent that the creatures were chiefly to be found in heathen climes.

In this passage on snakes, Derham offers two different kinds of teleology (the search for purpose): one anthropocentric — the notion that God has humans in mind in all that he makes or does, so snakes are perhaps a punishment — and the other focusing on function — the Aristotelian idea, even today a valuable guide to the natural historian, that animal form and behaviour always have some function, so snake venom must be useful to the snake. Derham is a powerful advocate of both ideas, and this gives his book its power. Linking science to religious concerns made it momentous. To some it seemed ridiculous, but really it was sublime as well as profound.

Since Babylonian times people had thought that the stars had influence, benign or baneful, over human destinies. The Christian church opposed astrology, as undermining free will. We could choose between good and evil, so our faults lay not in our stars but in ourselves. Nevertheless – partly in answer to the question, Why are there so many stars and planets? – many churchmen did admit that the stars affected us, even if they didn't control us. The sort of powers we attribute to our genes our ancestors down to the seventeenth century assigned to the stars.[12] But although, in the hands of skilled practitioners, astrology seemed to work – and when it didn't, one changed one's astrologer, or sympathised with his claim that more funding was needed for research – it ceased to be a reputable activity by 1700.[13]

For Derham the heavens (now much larger, as telescopes revealed countless stars never seen before) had no effect upon us except to make us wonder at the majesty and power of God who created them: they were awe-inspiring. We could only observe them, and marvel at the god-like Newton and his calculations of their motions; we could not experiment or dissect. Astronomy was a sublime science, but also a passive and contemplative one: suitable for Marys, while Marthas could busy themselves in laboratories.[14] The Providence revealed in astro-theology might be frowning, or certainly remote; in physico-theology we seemed to meet God taking care of his creatures. God was revealed in both the minute and the majestic.

Reason and uncertainty

Pagans had invented the argument for God from Design, Christians had christened it 'natural theology' – and the sceptic might wonder if any other theology was needed. William Wollaston, a Deist and first of a dynasty of scientists, in a book for which the young Benjamin Franklin set up the type, argued for a religion of nature.[15] Bacon had put his faith in two books, the Bible and Nature; Wollaston required only one: Nature. His ethics had no connection with the supposed will of God or the Ten Commandments: it was based upon a calculus of pains and pleasures. God, in creating the best of all possible worlds, would have used just such a calculus: so utilitarianism was another answer to the question why everyone was not happy all the time.

Our obligation as humans was, for Wollaston, to follow our reason. That would guide us to belief in God, whom we would be bound to worship; but the name Jesus does not appear in his index, and there seems no necessity for a special Revelation or for Atonement. God seems impersonal and remote: He has created an inscrutable world, where we are inevitably thwarted in seeking ultimate causes, and He cannot want or value the praise we are rightly disposed to offer. Reason will lead us into a prudent and sensible life and, if reason is unclear,

> that we ought to follow *probability*, when certainty leaves us, is plain: because then it becomes the *only* light and guide we have. For unless it is better to wander and fluctuate in *absolute* uncertainty than to follow such a guide; unless it be reasonable to put out our *candle*, because we have not the light of the *sun,* it must be reasonable to direct our steps by probability, when we have nothing clearer to walk by.

Science could lighten our darkness, even if it was only with a candle.

This idea of probability as the very guide to life was taken up by Bishop Joseph Butler in his *Analogy of Religion* (1736), a book which later delighted the Victorian statesman W.E. Gladstone. Butler, writing against Wollaston and other deists, emphasised the need for virtue rather than utilitarianism, the coherence of natural and revealed religion, and the incompleteness of natural religion on its own – though natural theology, as a partner of Scripture and church tradition, had a valuable place. His preface or 'advertisement' is famous:[16]

> It is come, I know not how, to be taken for granted, by many persons, that Christianity is not so much as a subject of inquiry; but that it is now, at length, discovered to be fictitious.

Where have we heard that recently? It seems there is nothing new under the sun, and predictions of the imminent collapse of religion are perhaps exaggerated. Butler's argument – that religion is a practical thing, and not just an intellectual one – did diminish the hostility to the argument from Design, which his more rigidly orthodox contemporaries saw as leading straight to Deism.

For Butler, arguments from analogy or probability were what we used all the time; and, if rightly applied, they indicated how well revealed and natural theology went together. Indeed, Butler's case for sensible reliance upon expert opinion, and for careful choice among probabilities, could be helpful when we have to make judgements today and look for more certainty from scientists than they can deliver.

Hume: the sceptic

David Hume was probably the most famous sceptic in eighteenth-century Britain. In his *Dialogues concerning Natural Religion*, posthumously published in 1779,[17] we find a classic exploration of the Design argument. The dialogue structure enabled him, as it had Plato, to put different sides of the question. He presented a dysteleological world – one without purpose – in which misery seemed to be magnified, where blind nature was 'pouring forth from her lap, without discernment or parental care, her maimed and abortive children'. As for our knowledge of the world, he argued that we have observed many houses and know about them so, when we see a house, we can infer a builder; but with worlds it is different. We only know one, and for all we know the past might be full of botched shots at worlds. Moreover, the Earth is 'cursed and polluted. A perpetual war is kindled amongst all living creatures'. Mankind is the prey of disease, superstition, fear and unhappiness. The 'course of nature tends not to human or animal felicity: Therefore it is not established for that purpose.'

Beautifully and wittily written, the book is very effective in demolishing the argument from Design. But we would be wrong to see Hume as a precursor of Charles Darwin. He was not presenting a scientific theory or even suggesting an alternative, but simply criticising physico-theology as a sceptic. Indeed, the orthodox could have agreed with much of what he said, as Gisborne did in his vision of the world as a reformatory, for the God of Nature is not easily reconciled with the God of the Bible. Even so, the cleverness of Hume was not going to undermine the religious-minded, whose faith did not depend on some calculus of pains and pleasures but on feeling sinful but redeemed. He was widely regarded with shock and horror.

Hume was not seen as a 'canonical' philosopher – to be studied by every student – until well into the nineteenth century, when a new breed of sceptics became interested in his remarks on natural theology, and also his ideas about causality: T.H. Huxley wrote a book about him. In Hume's own day, as he ruefully remarked, his philosophical writings dropped dead from the press; and he was mostly read as an historian of England.[18]

But William Paley, as we shall see, did engage with Hume's *Enquiry*; and Immanuel Kant, awoken from dogmatic slumber by reading Hume, realised that while the starry heavens and the conscience might turn our minds in wonder to God, none of the proofs of the existence of God stood up to logical scrutiny. Hume had been right. In the sciences, argument proceeds to a closure, at least in principle; or that anyway was how it seemed until the twentieth century. The Earth must be a perfect sphere, as Aristotle believed; or oblate, flattened at the Poles, as Newton predicted; or prolate, squeezed in at the Equator, as in Descartes' theory that it was spun around in a whirlpool or vortex of æther. Measurements made in the eighteenth century, by surveyors on expeditions sent out by the French Academy of Sciences, proved that Newton had been right. Objectivity, in the form of experimental or mathematical proof, is possible.

In the humanities, though, there is always something to be said on both sides of a question: and that was just what Hume had demonstrated in his *Dialogues*. The proof of a pudding is in the eating, and tastes differ. After Kant's explicit avowal of what was implicit in Humean scepticism, it was clear that persuasion and probability were involved in arguing for the existence, wisdom and benevolence of God. Preachers can very rarely have attempted a mathematical demonstration of God's existence – after all, a sermon would hardly have been the place for it or an effective way of changing anything – but philosophers and theologians sought proofs of various kinds, which are still logically interesting even if nobody thinks they have much to do with real religion.

William Paley

So it is that arguments in the humanities do not get settled; rhetoric rather than logic is crucial. One person who appreciated this was Paley,[19] whose name has become so associated with natural theology that people suppose that he invented it. His *Natural Theology* of 1802, published at the very end of a long life, is far and away his most famous book, but its author saw it only as one part of his campaign to place religion – and particularly the Church of England – on a firm base, in the face of attempts by deists like Wollaston, sceptics like Hume, and dissenters like Priestley to push the whole edifice over.

He was particularly encouraged by Shute Barrington, Bishop of Durham, who pressed him to bring out the book that would utterly confound atheism. Paley, a keen fisherman, replied 'My Lord, I shall work steadily at it when the fly-fishing season is over'.[20] He was not himself a man of science like Boyle, Ray or even Derham (who was an expert on wasps), though he associated with other clergy interested in natural philosophy. His great strength lay in his wide reading, and his ability to weave together ideas from authors ancient and modern to make a clear story. Like Priestley, he was said never to write an obscure sentence.

We should see Paley's *Natural Theology* as part of a strategy to show how the latest ideas fitted within Christian theology. This scheme also included his books on moral philosophy (where he was strongly utilitarian, like Wollaston and Priestley) and *Evidences of Christianity*, which sought to demonstrate how the prophecies of the Old Testament had been fulfilled in the New Dispensation. This was a rhetorical undertaking, presenting old ideas in new and acceptable clothing and engaging carefully with threatening notions. It was tricky, not to say dangerous. Benjamin Franklin had been converted to Deism by hearing a sermon preached against it; and we must all have been turned off voting for some political party by seeing their propaganda. Controversy must be carefully handled.

William Paley was a parson's son: born in 1743, he died in 1805. He went young up to Christ's College, Cambridge, and in 1763 was 'Senior Wrangler', meaning that he gained the highest honours of his year in mathematics. He next won a Latin prize, and duly became a Fellow and Tutor of his college. Then he married, and took a parish. In 1782 he was appointed Archdeacon of Carlisle. He ended his life at Bishop Wearmouth, now part of the industrial city of Sunderland but with a history going back to the Venerable Bede. This was a very well-endowed parish, bringing in a comfortable income, and he was also subdean of Lincoln, to which he paid an annual visit. He was in no hurry to publish. His writings came from the latter part of his life, and won him a Doctorate of Divinity – but not a deanery or bishopric, perhaps because he was seen as unsafe: too latitudinarian or liberal.

His sermons apparently failed to wind up with a concluding peroration, a mark of the most admired pulpit orators,[21] but he got his message across in print. His lectures at Cambridge were popular and he published them in 1785 as *The Principles of Moral and Political Philosophy*.[22] This was a great success: it became a standard textbook at Cambridge, widely read and admired. Making no pretensions to perfect originality, he claimed all the same to be more 'than a mere compiler': that claim is a feature of all his work. The book was dedicated to his friend and patron, Edmund Law, Bishop of Carlisle, whom he praised for demonstrating that 'whatever renders religion more rational, renders it more credible', thus purging it of ignorance and superstition.

Paley believed that previous writers had divided 'too much of the law of nature from the precepts of revelation'; as a good churchman, he aimed to keep the two in balance. Based on his experience that 'in discoursing to young minds of morality, it required more pains to make them perceive the difficulty, than to understand the solution', he began by exciting curiosity in order to arouse enthusiasm. The morality he sketched went along with the latest ideas, at the cost of being bland and simplistic. 'What promotes the public happiness' he wrote 'or happiness upon the whole, is agreeable to the fitness of things, to nature, to reason, and to truth' – and 'such is the divine character, that what promotes the general happiness is required by the will of God'. Hume's efforts to divorce morality from Christian theology had in the process enfeebled morality.

God's benevolence is shown, said Paley, in the way He has made our senses 'instruments of gratification and enjoyment'. Every smell might have been a stench, and every sound a discord – but in fact happiness has been promoted, and pains minimised. 'Contrivance proves design'. Actions are to be estimated by their 'tendency' – their outcome or utility – the very criterion he also used to infer God's benevolence. But we need also general rules, and to take into account the long run. The book contains powerful images of social injustice; and covers very readably a wide range of topics. It deserved to be successful.

Making sense of God's revelation

Paley's next book is generally said to be his most original. Published in 1790, it was a study of Saint Paul and the authenticity of his life and writings, called *Horae Paulinae*.[23] Paley makes an interesting point: if there were complete agreement between the epistles attributed to Saint Paul and the history in *The Acts of the Apostles* this would be evidence of 'meditation, artifice, and design' in the Bible. The harmony he looked for in establishing natural theology would weaken the case for independence here, and make it plausible that agreement had been concocted.

It is artlessness, or undesignedness, which is the sign of substantial truth here: and minute, circuitous and oblique circumstances, which no forger would ever have bothered with, are to be ferreted out in detective work:

WATCHMAKING

> If what is here offered shall add one thread to that complication of probabilities by which the Christian history is attested, the reader's attention will be repaid by the supreme importance of the subject; and my design will be fully answered.

Each chapter of the book examines a different epistle, looking for small links with the others and with *Acts* but also establishing as far as possible their independence as texts, and the unlikelihood that one was derived from another, or that several were the product of forgery. It is in Paley's attractively argumentative style: one long argument from beginning to end, indeed, and giving the impression that objections have been foreseen and considered, and are being dealt with in a fair-minded and judicious way.

In 1794 Paley published his *Evidences*,[24] bringing him great fame as an apologist: here as elsewhere he was looking for coherence. This book was much studied at Cambridge, and in the 1830s at the infant University of Durham it was the basis for the compulsory course on Rudiments of Religion. It is concerned with the Biblical miracles, and with the historical evidence for Christianity: and thus takes on Hume, Gibbon, and earlier Deists like John Toland for whom true Christianity was not mysterious. Paley considers prophecy, the morality of the Gospels, the character of Jesus and the history of the Resurrection. Having reflected upon these things, he considers the success of Islam, 'Mahometanism', generally seen by Christian apologists in the past as an imposture. Paley was no different: Islam could not be compared with Christianity, because it was essentially military and political and its rapid spread was by conquest.

Then, looking as usual for objections, he examines the discrepancies between the Gospels and other problems that confronted expositors. The book clearly did its job in the fraught atmosphere of the 1790s, appearing in the very year of Robespierre's Reign of Terror and the British reaction to it; and it stayed in print for many years. What was now needed to complete Paley's project was to take up and treat fully the argument from Design touched on in his earlier books.

Very often writers, looking for the classic statement of the idea that every animal and plant was created by God to fill the place it inhabits, come backwards from Darwin to Paley. And because the *Natural Theology* was published in 1802, they treat it as a text for the new century. Paley indeed kept up to date as far as possible with his reading, but he was essentially a man of the eighteenth century; and it is better to see the book as the last component of a great engine designed to crush atheism and not as something seminal for the next century. Nevertheless, there are ways in which it was a model for Darwin's *Origin of Species* of 1859, and we should be aware of that.

Taking the back off God's chronometer

Paley's *Natural Theology* begins[25] with a much-celebrated and anthologised first chapter, about a watch. If we were to come across a watch when out for a walk, on picking it up we would note how the hands (visible through the glass) point to the time, and if we opened the back we would see how all the springs and gear wheels work together, some brass and some steel. It would be absurd to suppose that the watch was some chance congeries of atoms: it is an artefact and must have had a designer and maker. If we were to find that it contained mechanisms for generating further watches, our respect for its artificer would be all the greater. The world is a great watch, and we little watches; the atheist is the ninny who denies that watches, worlds, humans and animals are designed.

It was one thing for Boyle to be bowled over by the Strasbourg clock: such intricate machinery was still rare and novel in the seventeenth century. When he had advanced some distance up the ladder in the Navy Office, and it was appropriate to his new and grand status, Boyle's younger contemporary (and future president of the Royal Society) Samuel Pepys[26] accepted a present of a watch from Mr Briggs, with whom he did business; and was gratified to find it was worth £14. Clumsy turnip-like contraptions that kept rather poor time, they were the summit of high technology and miniaturisation towards the end of Boyle's life.

By the eighteenth century, they had become relatively commonplace. A gentleman would expect to have a watch, with a chain perhaps extending across his good round belly; they were no longer an innovation. Further, the 'mechanical world-view' of Boyle and his contemporaries was in eclipse in the English-speaking world, because of those political connections with absolutism: the United States' Constitution was based on simpler systems of checks and balances – though the best clockwork had those too.

It might therefore seem surprising that Paley should have chosen a watch in making out a new version of the Design argument, when he might perhaps have had a steam-engine as the latest thing. There are two possible reasons why he didn't: one was the very familiarity of watches. The illustration was in 1802 a homely one, based on an analogy that all readers would recognise; and in general, Paley used established knowledge in the book even while discussing the very latest ideas – perhaps he was chary of trendy notions and devices.

The other possibility is that the invention of the chronometer by John Harrison[27] meant that the old analogy could be resurrected: watches had achieved astonishing standards of accuracy. Captain James Cook had taken chronometers on his second voyage into the Pacific, in 1772–5, making his navigation much easier and more accurate;[28] and by 1802 one of the major ongoing activities at Greenwich Observatory was the checking or 'rating' of chronometers.

Beyond reasonable doubt

Paley had earlier alluded to 'that complication of probabilities' by which Christianity was supported. His study of Saint Paul had been lawyer-like: he applied forensic ingenuity to the texts to establish that they were too consistent in small and unexpected details to be fictitious – a telling point. In our generation we know, from careful studies of Shakespeare's plays or Victorian novels,[29] how difficult it is for an author to compose a completely coherent story.

Paley knew – as his contemporaries did, even before Hume and Kant pointed it out formally – that God's existence cannot be proved in the same kind of way as Pythagoras' theorem or the complex reasonings of Archimedes[30] in geometry. If readers have accepted the postulates and axioms – common notions that seem self-evident (such as, if A is bigger than B, and B is bigger than C, then A is bigger than C) – they are inexorably led towards surprising conclusions about the square on the hypotenuse or the properties of conic sections.

Religion, and indeed ordinary life, is nothing like that: we are humans and not calculating machines, and we have to carry on in the real world, where probability is the guide. Proof in law courts is nothing like mathematical proof: it means that the case is made beyond reasonable doubt, and we have appeal processes just because judges and juries are fallible in their judgements, and evidence is never complete.

Inferring a Designer is not something upon which everyone will agree. Some people are Doubting Thomases, sceptical about what seems knock-down evidence; others are very ready to accept well-constructed arguments. Anyone presented with the argument from apparent design and contrivance to the existence of God will naturally want to think about it and test it, so it is appropriate and proper to look at those who, like Paley, did just that but came to opposite conclusions. Natural theologians include seekers as well as finders.

Geometrical theorems are a good example of purely deductive thinking, pulling out conclusions that are implicit in the axioms but nevertheless are sometimes startling. Geometrical proofs have to be taken in the right order: Pythagoras' theorem depends upon a series of other deductions that came first in the book. It was Euclid's great achievement[31] to order the theorems so that separate results were incorporated into a great structure of formal proof; and mathematicians spend a lot of time rigorously proving what nobody doubts. We get from the axioms to the final theorem by way of a chain of other proofs; and a chain is only as strong as its weakest link. If therefore any of the theorems on the way to our conclusion is not proven, the argument collapses.

In ordinary informal life, and in the argument from design, it is different. The preacher, the politician, the advocate, and the wooer present a series of arguments for seeing things their way, but while these are convergent they are also logically independent. Manifestos are not like Euclid: we may reject a number of the things a party stands for, and yet vote for its candidate – we may likewise find some of Paley's points unconvincing (as contemporaries did) and yet assent to his general conclusion. Such arguments are not like a chain, but like a rope; some strands may break under strain but, well twisted, the rope may nevertheless bear the load put upon it. Such arguments are not ropy, they are just informal by comparison with deductions.

In arguing for God's existence, Paley also had to argue for His wisdom and benevolence; for his was the God of Abraham, Isaac and Jacob, of Jesus and the apostles, and not some impersonal First Cause or remote Deity. It would have been just as bad for him if a reader ended up a Deist, or remained a sceptic. We met in his earlier writing the idea that God had made pleasures from what might have just been refuelling: we enjoy our food. Indeed, for Paley even the shrimps swimming in the still relatively unpolluted North Sea – and living, like other animals, only in the present – were happy.[32] Hume's melancholy vision was untrue to life; Nature took good care of her progeny, even planning ahead. We find prospective contrivance: God has put into us and other organisms structures, like our second teeth, that we do not need at first, but which will develop when the time is ripe.

Undoubtedly, there is pain and misery in the world; but it is inevitable, and is minimised. Predators can be seen as beneficent:[33] the ageing antelope is spared the pains of old age, it is all over very quickly, and the wolf or tiger and its family benefit from a good meaty meal. On a utilitarian calculus, there is a gain all round. Instead of being looked at with horror, hyenas, crocodiles and such creatures should be seen as euthanasia machines. This is a poor comfort to the widow or the orphan of somebody who has been eaten by one; and indeed the Design argument usually comes from the healthy and comfortably off. Chillingly, if this is indeed the best of all possible worlds, then there is nothing to be done about its evils.

God's handiwork

Paley's book is a work of rhetoric. It begins with a statement of the argument, the chapters about the watch; and then continues to examine in detail the evidence, which all converges upon the conclusion that the world too is the product of Design, benevolent and far-seeing. As the cumulative argument builds up with example after example, the reader finds it hard to resist, carried along by Paley's easy but knowledgeable style. A great deal of natural history is effortlessly acquired on the way: the book (and others in its genre) should never be underrated as excellent popular science, engaging well with the reader, who is being constantly educated but never patronised.

The story begins with the eye, in humans, fish and other species; and continues with the mechanical arrangements in the body. This is shown to be full of mechanical contrivances, for example at the elbow joint. The spine similarly is a wonderful multi-purpose construction. The complex and efficient arrangements of the throat and larynx are amazing: 'not two guests are choked in a century' by crumbs going down the wrong way.[34] The diversity of organisms, and yet their unity, is stressed; so we get interesting reports of how caterpillars cast off their teeth in the chrysalis and emerge with quite a different kind of mouth; and an account of the uses to which the woodpecker puts its tongue. Then, the various parts are all related together so as to fit the life the creature leads; and this can even, as with the mole's eyes, mean atrophy. Apparent deficiencies in one part of an animal are found to be compensated for by advantages elsewhere.

The second half of the book relates animals and plants to the world around them. There is a chapter on the solar system and Buffon's hypothesis of its evolution (which is trenchantly criticised), but this is not a book of astro-theology; Paley did not believe that astronomy was the best evidence for the agency of an intelligent Creator, though it did display the magnificence of His operations.[35] Paley is in Ray's tradition. The wings of birds and the fins of fishes are adapted to the media in which they pass their lives, air and water; as our ears and voices are similarly adjusted to the air, and our eyes to the light. Natural law, or some notion of development, could not account for all this order. Then came the crucial part, where Paley argued for the personality, attributes and goodness of God. Referring to Priestley, Paley argues that 'contrivance' proves design, and that which can design must be a person: 'The acts of a mind prove the existence of a mind; and in whatever mind resides, is a person. The seat of intellect is a person.' Moreover, living, active, moving, productive nature is evidence of continuing action, of God sustaining the universe.

Adaptation, uniformity of plan, the absurd improbability of evolutionary systems, and the pleasurable and beneficial effects of contrivance make the points about God's nature: 'Nor is the design abortive. It is a happy world after all. The air, the earth, the water, teem with delighted existence. In a spring noon, or a summer evening, on whichever side I turn my eyes, myriads of happy beings crowd upon my view.' But he recognised also the 'superfecundity' of nature, the way in which, unchecked, the progeny of animals would over-run the Earth: and explained it in terms of defencelessness and devastation being repaired, thanks to God's foresight, by fecundity. Apparent evil and chance have much to do with our ignorance, and are trials of character.

The conclusion was that natural theology was valuable because it turned our mind constantly to God, increasing the stability and impression of our faith. And it prepared the mind for revelation, meeting God through His word and through His church. Despite the order in which his books had appeared, the *Natural Theology* was in effect the preparation for the *Evidences*; and perhaps both were preparation for the close sort of study of the Bible set out in *Horae Paulinae*, and then for the ethical reasoning in *Moral Philosophy*.

Although the *Natural Theology* and the *Evidences* were enormously successful, they were not something that appealed to everyone. A religion based upon evidences, with juror-like weighing of probabilities and a utilitarian ethic, did not seem to be what the apostles preached; nor was it likely they thought of Christianity as revelation built upon a foundation of natural theology. And like any rhetorical piece, Paley's was not a completely knock-down argument. Indeed, when *Natural Theology* appeared, Erasmus Darwin (an author whom Paley warily quoted) was writing *The Temple of Nature*, published posthumously in 1803,[36] urging a theory of progressive evolution that the next generation would refer to as Darwinian – meaning Erasmus, not Charles.

Erasmus Darwin,[37] though an older contemporary of Paley, seems to us more modern in his emphasis on real change over time – a genuine historical awareness, carried back into prehistory and the very beginning of the world. Paley's world was a timeless one: Newton's time indeed flowed on, an absolute framework into which events could be slotted as they had been in Priestley's time charts. But if this was the best of all possible worlds, change could only be for the worse; and significant change was therefore not to be expected. Nevertheless, in the early 1800s discussions of providence were complicated by awareness of very significant changes. There had once been creatures that existed no longer: what was God's purpose in allowing them to become extinct?

True religion is not reasonable

Darwin was a freethinker, in the tradition of eighteenth-century scepticism; but there were also good Christians who found Paley's reasonings distasteful. His was a God apparently careless of the single life, devoted instead to a utilitarian calculus. And the search for evidences for religion began to seem misconceived. Believing in God was not like believing in waves or particles of light, billiard-ball atoms, the electric fluid, or phlogiston – theoretical entities prominent in the science of about 1800. It involved trust, like believing in one's spouse or country – a different kind of belief, where nobody doubted their existence but only how wise it might be to count on them.

The leap of faith was stressed in Methodism, but also in Germany by Friedrich Schleiermacher in 1799 in his *Lectures on Religion to its Cultured Despisers*,[38] given in Berlin and subsequently published. He was facing the rationalistic writings of G.E. Lessing[39] and the critical philosophy of Kant and his disciples, especially J.G. Fichte.[40] Fichte had asked how we would judge whether a prophet was true or false; and concluded that we would judge any claimed revelation of God against the standard of natural religion and morality. Natural religion was therefore more than the basis for revealed religion: it was also its criterion. Why then was the ministry of Jesus necessary? Fichte could only conclude that religion was at such a low ebb at that time that a major recalling of mankind was required. Jesus was in effect a revivalist. Such minimal doctrine was clearly a long way from the beliefs of practising Christians.

For Schleiermacher, coming from a tradition of pietism, religion was not to do with moralists and metaphysicians, who believed that 'no drop of religion can be mixed with ethical life without, as it were, phlogisticating it and robbing it of its purity'.[41] They completely failed to comprehend how 'When the world spirit has majestically revealed itself to us, when we have overheard its action guided by such magnificently conceived and excellent laws, what is more natural than to be permeated by a heartfelt reverence in the face of the eternal and invisible?' Feelings of awe, gratitude and common humanity – not dry and elaborate reasoning – were the key to true religion.

Disillusioned by the French Revolution and its exposure of the emptiness of Enlightenment, Schleiermacher appealed for genuine and particular historical religion rather than the thin and dry faith of Deists, or the fence-sitting of sceptics:

> So-called natural religion is usually so refined and has such philosophical and moral manners that it allows little of the unique character of religion to shine through; it knows how to live so politely, to restrain and accommodate itself so well, that it is tolerated everywhere. In contrast, every positive religion has exceedingly strong features and a very marked physiognomy, so that it unfailingly reminds one of what it really is with every movement it makes and with every glance one casts upon it. . . . If you realize that a special and noble human capacity lies at its core, which must consequently also be cultivated wherever it shows itself, then it cannot be offensive to you to intuit it in the determinate forms in which it has already actually appeared. Rather, you must deem these all the more worthy of your contemplation, the more the unique and distinguishing features of religion are formed in them.

Those who took religion seriously would therefore have little time either for the sneers of sceptics or for the elaborate search for evidences that had preoccupied Paley. To congratulate God upon his ingenious design of a shrimp or flea, or to vindicate His actions in accordance with utilitarian calculations, was absurd or blasphemous.

The nineteenth century thus opened with two major publications. Paley's was a classic statement of the argument from Design, cast in the traditional and timeless mode of the Enlightenment but designed to refute Deists and sceptics and to prepare the way for the acceptance of revealed Christianity, whose doctrines would be found to be in harmony with those of the natural theology which underlay them but which went deeper. Schleiermacher would have none of that. For him, religion was not reasonable. It had always affronted people. It was not based upon the careful weighing of evidence, but upon a leap of faith, and inner conviction; and necessarily included historical particulars, and traditions – Gibbon's 'barbarism and superstition'.

Paley is a figure of the Enlightenment, Schleiermacher a Romantic. For those who followed Paley, scientific investigations had religious significance: they were revealing God through His works. For disciples of Schleiermacher, God was inscrutable in that way: science could be important for making life easier, and the excitement of discovery might be sublime — but it was not momentous, as Paley thought. To see the world as a great clock and God as the Clockmaker (probably still enjoying His Sabbath-day rest while his mechanism ground steadily on) was repugnant. These years were also years of tremendous scientific change, which required in the next generation a rethinking and restatement of the relationship between science and religion, once peace had again returned to Europe and the world in 1815.

CHAPTER 4

Wisdom and benevolence

THE YEARS AROUND 1800 WERE crucial ones for the development of modern science. Previous generations – Galileo, Descartes, Boyle, Newton and their contemporaries – had made scientific thinking seem a distinct and impressive kind of thinking. They had also focused attention on nature as a mechanism, and God as its contriver, displaying indeed His wisdom and benevolence thereby. Newton had bequeathed to his successors the ideal of seeking the simple mathematical relationships that were now seen to underlie what had previously seemed complex and disorderly phenomena. In the English-speaking world, though, his astronomical work seemed to have been essentially completed. In his *Opticks* of 1704 and later editions[1] Newton made intriguing suggestions not only about the nature of light but also about matter, about method, and about what kind of world it was. These 'queries' were to stimulate thought and debate for over a hundred years, and still make fascinating reading.

There was an alternative, though, to both the applied mathematics of Newton's *Principia Mathematica* of 1687[2] and his experimental *Opticks*: the descriptive science of Charles Linnaeus,[3] the great botanist whose system of binomial nomenclature names brought order into natural history, just at the time when sailors such as James Cook were making careful exploration possible (through the conquest of scurvy, and success in finding longitude) and demonstrating how varied the world was. Newton and Linnaeus, mechanism and classification, represent the Enlightenment promise of the power of science to transform the way we contemplate and manipulate the world. Both kinds of science involved close and careful empirical work.

At the end of the eighteenth century this orderly world was threatened by unexpected change: the French Revolution, but also the Romantic movement[4] with the belief that there are more (and more interesting) things in heaven and earth than in the tidy dreams of natural philosophers. These great upsets were associated with changes in the way science was organised: careers in science became possible, scientific societies and institutions proliferated, and specialisation began in a more serious way. People began to perceive the relationship between science and religion differently. By 1830 it was time for a major restatement, which became a great publishing success – the Bridgewater Treatises.

The status of science

To make a career in science was problematic. In France there had been since the 1660s an Academy of Sciences, with a limited, salaried membership chosen by the king (on the recommendation of existing academicians, once it was set going).[5] This model was imitated in the eighteenth century in Prussia, Russia and other countries, and there were thus a few men who could live by their science. Their income would not match that of a fashionable doctor; but their social position, and the authority it brought, made a career in science a tempting prospect for an able boy with inclinations that way.

The problem was the limit on numbers. Somehow the candidate had to make a reputation for himself so that when an academician died he would be the obvious person to step into those shoes. As there were a fixed (and small) number of places for chemists, astronomers and so on, and academicians might drop dead at any time or survive into ripe old age, the whole process was very uncertain – Bernard le B. de Fontenelle (1657–1757) lived up to his title of Permanent Secretary of the Parisian Academy of Sciences by going on into his hundredth year. Ambitious young men faced an indefinite wait, doing their best to cultivate the academicians who might nominate them if and when the time came. In Britain and North America, there was no such possibility of a career based upon scientific research; the only post was that of Astronomer Royal, directing Greenwich Observatory and having particular responsibility for the improvement of navigation.

Everywhere, however, medicine – and medical schools – provided one good opportunity for scientific life. Remedies available in the late eighteenth century included opium, quinine (in the form of 'Jesuits' bark') and alcohol;[6] and it was very difficult to be sure whether even these really did very much good, in the absence of statistical methods and pure samples. A doctor's reputation depended on his bedside manner and prognoses – predictions of what would happen. Doctors nowadays hate telling patients they are dying, but then it was recognised as an essential part of the job, so that the family, the lawyer and the parson could be summoned in time to the deathbed. And what distinguished the real doctor from the quack in the medical free market was his scientific knowledge, which also brought status.

Some eminent doctors not only assembled clinical experience but also did important scientific research, in their free time or perhaps in conjunction with teaching, often in anatomy and physiology but also in chemistry, natural history, and even mechanics and geology. But though there were in Leyden, Edinburgh and elsewhere professors doing original science, during the eighteenth century research was seen as an optional extra for a university professor, who was expected (in medicine as in classics) to teach a traditional syllabus.

Only in a few institutions such as the new University of Göttingen was there a perceived need to advance knowledge and make discoveries.

Lawyers might also find some leisure for scientific investigation, and were particularly prominent in the Edinburgh enlightenment of the later eighteenth century. So also might officers in the army and the navy – and by the late eighteenth century naval surgeons counted as officers rather than NCOs, while trimming hair and beards ceased to be an important part of their duties. For our purposes, however, the clergy are especially important as a profession where interest in expanding our knowledge of the world was seen as appropriate.[7] We all know about Gilbert White of Selborne, and his careful recordings of the wildlife and farming of his village – and also, as became a learned man, of its antiquities and history. He was not unusual, and incumbents in his mould were to be found in parishes throughout Britain – Scotland and Wales as well as England – in the nineteenth century and beyond. Dissenters too, especially Unitarian ministers committed to religion within the limits of reason and to the argument from design, were active in science and encouraged their flocks to pursue it.

During the nineteenth century, however, ideas about what was appropriate for clergy narrowed. In 1854 Samuel Wilberforce – 'Soapy Sam' to unadmiring contemporaries – was one of the first English bishops to found a theological college or seminary (near his own residence as Bishop of Oxford, at Cuddesdon) to train ordinands. Candidates had hitherto just done a little recommended reading after taking a degree in classics or mathematics. Wilberforce also pursued the parsons in his diocese with hard questions about exactly how they were carrying out their duties. This process continued to the point where, at the start of the twenty-first century, clergy of all denominations are very busy and professionally conscious, with little time or energy for activities outside their profession – narrowly defined.

Organised science

Newton's successor at Cambridge, William Whiston, had been sacked for noisily propagating the Arian heresy. He took up public lecturing, and this provided him and others with a living. Such lecturers[8] were often itinerant, like John Wesley, going from place to place and arousing enthusiasm – in their case for science. As a rule, the science centred on Nature and Nature's God, for natural theology was a good sweetener, and was also generally and genuinely accepted.

They would often demonstrate an air pump, optical, mechanical and chemical experiments, and perhaps an electrical machine; and they might well write a book to accompany their lectures for everyone knows that, while a lecture is good for raising interest and conveying personal commitment, it is not efficient for getting information across. Such private arrangements were in the last decade of the eighteenth century complemented by new institutions being set up, where lecturers might hope for something more like regular employment. Together with publishing in journals and books, lecturing opened the possibility of a scientific career in the English-speaking world. These societies opened up science for new publics, and also brought together specialists.

There was the Royal Society in London, and corresponding bodies (akin to gentlemen's clubs) were set up in Dublin and Edinburgh; while, from 1743 in Philadelphia and 1779 in Boston, America had its learned societies too. There were also less formal gatherings in coffee-houses.[9] The Literary and Philosophical Societies set up in the industrial cities of Manchester and then Newcastle-upon-Tyne were rather different from the metropolitan clubs, because their members' social backgrounds were so different: these were cities where ingenuity and skill were beginning to pay off in new activities.[10] Unitarian clergy, and sometimes also more orthodox ones, and medical men[11] (usually surgeons rather than graduate physicians) were prominent in founding such societies: and the industrialists who joined were often very conscious of their lack of formal education and polish.

One might have supposed that such societies would have devoted their efforts to applied science, as the select but informal Lunar Society of Birmingham[12] had tended to do, but it seems that the last thing manufacturers wanted to do was to give away trade secrets to rivals. Instead they wanted to meet for intellectual pleasure, to relax by hearing about natural history or some arcane discoveries in the sublime science of astronomy or the rapidly developing field of chemistry.

The Manchester society, established in the raw boom town, was lucky to attract John Dalton the Quaker as its anchor man; on a small salary eked out by giving private lessons, he looked after it and its publications for about half a century. Such societies did not confine themselves, in these days before 'two cultures', to what later generations would call science. Literature and 'philosophy' – which then, like 'science', meant any body of established and organised knowledge – these were their brief, and we can best think of them as devoted to establishing and promoting high culture in their cities. They were followed in 1799 by the Royal Institution (where Humphry Davy, Michael Faraday and John Tyndall were to work) and other places in London, and also in provincial cities. What they provided was programme of lectures, a library, and perhaps a journal and a post for a man of science, who could do research somewhere on the premises and might live, as it were, above the shop.

Societies such as these were the beginning of what contemporaries called the March of Mind (or Intellect), which by the 1820s was being spread ever further by cheaper books and the expansion of church elementary schools. Cheaper printing, using steam presses and wood-pulp paper, brought a sharp drop in the price of books. In the quarter-century after 1828, the average price of a book in Britain fell by 48 per cent, from 16/- to 8/4½, a development in which Henry Bohn was especially important.[13]

The middle-class Literary and Philosophical Societies, or Athenaeums, were joined by Mechanics' Institutes, whose membership was drawn from skilled members of the working class.[14] These might be run by the members, or for them by some philanthropic employer. Their libraries were very important, giving people who would previously have been cut off from written culture the chance to read books. They also had courses of lectures, and self-improvement seminars where the members took it in turns each week to give a talk on some intellectual topic or perhaps to demonstrate an experiment.

Even in England, education was coming to be seen as a necessity, a key to progress rather than a privilege. In the 1820s the self-styled secular 'London University' was founded to provide higher education without religious tests:[15] although it was denounced as Godless, many of its students were in fact members of religious bodies of one kind or another. Church of England rivals in London and Durham were being planned, and had good prospects of being formally chartered and established.

Associated with the March of Mind was the expectation that science was the key to a better life, exemplified in technical progress like steam engines and Davy's safety lamp for coal miners. By 1830 scientific culture was far more widely diffused than it had been in 1790, and was no longer chiefly confined to London and ancient university towns. There was a new market for a new natural theology.

The specialists

All the institutions so far mentioned covered the whole domain of natural philosophy and natural history, often with literature as well. Their journals included articles on botany, medical sciences, geology, chemistry, physics, astronomy and mathematics; and all these disciplines and more featured in their lecture programmes. By contrast, the Germans in the later eighteenth century had had a journal devoted solely to chemistry,[16] and Lavoisier and his associates launched another in France.[17] In London in 1788 the Linnean Society was founded, to house Linnaeus' collections (bought from his heirs by the banker James Edward Smith) and to promote natural history. Its meetings and journal were confined to this field.

Sir Joseph Banks had welcomed and been associated with this project; but with the coming of the French Revolution and war he rapidly went off the idea of specialised societies. Believing that if men of science did not hang together they would hang separately, he saw the Royal Society as the centre for all science, within which subgroups of specialists (like the Animal Chemistry Club in the early nineteenth century) might happily get together. This naturally also meant that all science would be part of his learned empire; he was not a man to favour breakaways. In any case, with a small scientific community the risks of fragmentation were real, and Banks' opposition should not be put down (as it sometimes was) to megalomania alone.

The Geological Society was founded in 1807, to Banks' fury. He made his protégé Davy resign from it, but despite Banksian opposition it flourished. It met a need. Its meetings were also very lively compared to those of the Royal Society: there was real debate. The science was going through a period of tremendous change, with fossils being reinterpreted as the remains of creatures that had inhabited one or more 'former worlds'. Georges Cuvier in France was the most important person involved in this work,[18] but in Britain there were also many involved in mining, canal digging and agriculture who, with university men and collectors, could coalesce into an active society that would describe rocks and begin to date them using the fossils they contained.

When Banks died, Davy as his successor was prepared to countenance specialised societies, dedicated for example to astronomy and zoology: science was inevitably breaking up into sciences. Alexander von Humboldt was working across a whole range of sciences, but he was a prodigy: Sir John Herschel deliberately decided not to specialise, but even then he could really only cover physics. To make a career it was essential to become an expert; and so men of science naturally tended to go to the appropriate specialised learned society to report their research and meet their friends and rivals.

This meant that, outside their specialism, men of science were laymen and needed to be kept up to date. The Royal Institution or their local 'Lit & Phil' might do the job very well, as did Mary Somerville with her books on science directed to the well-educated who wanted an overview. Thus by the 1820s books on natural theology also began to specialise, focusing upon one science, though they might also fit into a general scheme of getting special knowledge across to wider publics. Paley's *Natural Theology* was often reissued, revised and illustrated, but by the 1820s sciences had so proliferated and science so changed that something new and more closely focused upon single sciences was needed, even if the basic arguments were little different from Paley's. Such writings helped readers to place specialised knowledge in context, relating it to other facts but also to a world-view with intellectual and moral aspects.

The Bridgewater Treatises

In February 1829 the Earl of Bridgewater, heir to an enormous fortune from canals and mines around Manchester, died. He was a clergyman, holding a canonry of Durham Cathedral; but his duties had been light, enabling him to live most of the time in Paris, surrounded by cats and dogs who ate with him at table. In 1825 he had made his will, bequeathing the substantial sum of £8,000 to be applied by the president of the Royal Society as payment to one or more authors whom he was to select, who would write and publish a thousand copies of a book

> On the Power, Wisdom, and Goodness of God, as manifested in the Creation; illustrating such work by all reasonable arguments, as for

instance the variety and formation of God's creatures in the animal, vegetable and mineral kingdoms; the effect of digestion, and thereby of conversion; the construction of the hand of man, and an infinite variety of other arguments; as also by discoveries ancient and modern, in arts, sciences, and the whole extent of literature.

The profits from the book(s) were to go to the author(s).

Davy had resigned as president, and his successor Davies Gilbert consulted with the Archbishop of Canterbury, the ultra-Tory William Howley, with the more pragmatic and bustling Bishop of London, Charles Blomfield,[19] and with a member of Lord Bridgewater's family; and the decision was taken to nominate eight authors to cover the various branches of science. This was arithmetically convenient, but even so it seemed that eight books were needed in 1829 to do what earlier authors had achieved in one or two volumes: this was a period of very rapid growth and transformation in the sciences.

There was some difficulty about finding a publisher:[20] those first approached doubted that such a series would go down well, and in the end the agreement was made with the small publisher of high-quality books, William Pickering. In fact the series was an immense success. It met a real need for high-level popular science, fitted into a spiritual framework that gave it deep significance by linking science with other concerns.

The extract from the will, and the list of the authors chosen, appeared at the front of each book; so that it was clear to all that they formed a set, even though they were not all published at once, or in order, but over a few years as the authors finished their task — Gilbert's successor as President of the Royal Society, the Duke of Sussex, had desired that there should be no unnecessary delay in getting the message across in this time of agitation and reform. Many were bought as sets, and often rebound in leather to look well on the shelf. Among would-be authors who were not chosen there were some grumblers, who murmured that arguments for the existence and goodness of God were less convincing once you knew the author had been paid £1,000 to make them. That sum was about the maximum annual income Sir Humphry Davy had attained as professor at the Royal Institution and from other sources, before his marriage to a wealthy widow.

The Bridgewater authors

The first author on the list was Thomas Chalmers, a minister of the (Presbyterian) Church of Scotland famous for his sermons on Newtonian astronomy, writing about how external nature was adapted to our moral and intellectual constitution — most of us nowadays would see this the other way round. The rest were Anglicans, three of them clergy (though only one was in ordinary parish work) and four medical men. John Kidd was Professor of Medicine in Oxford, and wrote about how external nature was adapted to our physical condition (again, this seems upside down in our day); and Peter Mark Roget, Secretary of the Royal Society, now famous for his *Thesaurus* familiar to devotees of crossword puzzles, wrote about animal and vegetable physiology.

William Prout (like Roget, an eminent London practitioner)[21] had a rag-bag of a treatise, covering chemistry, meteorology and the function of digestion. Chemistry is a tricky science because it is so concerned with improving the world rather then contemplating or admiring it:[22] Prout's tome is most interesting for its atomic and molecular theory – he is famous for his hypothesis that all the chemical elements were polymers of hydrogen – and for its passages on digestion, for he was the discoverer of hydrochloric acid in the stomach. Meteorology was of great contemporary interest, for Luke Howard[23] had begun the classification of clouds, Captain Robert FitzRoy was using barometers to predict gales and williwaws on HMS *Beagle* with Charles Darwin, and John Constable and Joseph Mallord Turner were making cloud studies for their paintings. While excellent in parts, Prout's volume could hardly be said to be one long argument.

The fourth doctor was Charles Bell, Professor of Surgery in Edinburgh and eminent as the man who had distinguished the endings of motor and sensory nerves while eschewing the vivisection used by his French rival, 'the murderous Magendie'. He was also famous as an artist[24] and connoisseur, being especially interested in how our facial muscles enable us to be expressive. Charles Darwin later wrote a book, *The Expression of the Emotions in Men and Animals* (1872) as an attempted refutation of Bell's argument in his *Anatomy and Philosophy of Expression* (1806) that we had been designed to communicate with each other. Because Lord Bridgewater had referred explicitly to the hand in his will – perhaps as a result of watching those cats and dogs eating – Bell undertook a whole volume devoted to it as an example of design.[25] It contains a good deal of comparative anatomy, showing how our hands resemble and differ from the walrus's flipper and the horse's hoof. There is also an interesting discussion of 'phantom limbs' in which patients feel sensations after amputations.[26]

The book is attractive, and shows Bell's wide learning and broad sympathies – he was no narrow specialist. The focus upon a single organ, while allowing for digressions, gives a coherence to the volume: authors lacking Paley's rhetorical skill could easily find that their books became a kind of shapeless piling up of instances. Any reader of Bell's book would have got a clear impression of what anatomists were up to as well as a strong feeling for the structure of animals and their parts, and the importance of our hands.

There is a temptation to see the Bridgewater authors as a team of liberal Anglicans with a party line: but that is a mistake. Chalmers was a Presbyterian; and the others, whether clergy or laymen, differed as members of broad churches do. The other three clergymen were William Buckland – who, as a geologist, warrants a separate chapter in this book – William Kirby and William Whewell. All of them were deeply involved in science, Buckland and Whewell holding chairs at Oxford and at Cambridge.

Kirby, vicar of Barham in Suffolk from 1782 to his death in 1850, was an old-fashioned high-churchman – born in 1759 – in the tradition of William Jones of Nayland. He was a founder-member of the Linnean Society who made the study of insects popular in the delightful book he wrote with William Spence, *An Introduction to Entomology* (4 volumes, 1815–26) which began – like White's *Selborne* – in the form of letters, turning in volumes 3 and 4 into something a little more formal, but still admirably personal and discursive.

The *Introduction* contains a particularly interesting discussion of instinct – obviously an important question to anybody interested in design. The book was a success, with several editions, and combated the view that studying insects (which were only just being separated from other kinds of creepy-crawlies) was ridiculous. In 1818 Kirby was elected a Fellow of the Royal Society. Because of his theological tradition and his base in a country parish, Kirby's Bridgewater text is more suffused with spirituality than most of the others: natural theology can readily go with rationalism akin to Deism, but he was strongly averse to that. References to the Creator are not just bolted on to an otherwise straightforward discussion of nature. Kirby's was a sacramental faith.

Kirby on animals

Whereas a rather literal reading of the Bible, as though it was a plain text, was becoming the norm in the eighteenth and early nineteenth centuries, Kirby belonged to an older tradition in which allegorical readings were also expected and indeed emphasised. Kirby's book is in many ways the oddest of the Bridgewater series.[27] His subject was the creation, history, habits and instincts of animals; he chose, not surprisingly, to focus frequently upon invertebrates, though the title on the spine of the 1853 edition[28] was simply 'Kirby on Animals'.

Mankind, he concluded, was unique in belonging to both the material and the immaterial world: our divine image and dominion over other creatures were damaged at the Fall of Adam and Eve, but not abrogated. While some animals had escaped from their servitude, many had not, and the wild ones still feared us and fled. Man had then had to arm himself to reduce the number of predators, and to destroy rebellious creatures; and this warfare (by his time, the gentry keeping down vermin) had inexorably led to war between men.

In Kirby's book, as well as putting animals in their Linnean and social place, we could learn spiritual lessons from studying them, in particular from 'representation' and its associated symbolism. Minerals grow, representing trees and plants; insects and plants resemble each other, and the zoophytes (animal/plants, creatures like corals) have been named accordingly. As we ascend the scale of nature, so we find that everything has its representative somewhere: and the parallels extend from form to character, rapacious or tractable. Naturalists of the time, for example, described how the kangaroo 'represents' the deer in Australia, filling a similar niche.

Kirby went further: our great Instructor had made animals thus for a purpose, placing before us an open though mystical book, giving us instruction about the spiritual world. What is symbolically revealed is that just as animals are benevolent and beneficent, or malevolent and mischievous, so are spirits – angels or devils. For him, the language of symbols and types pervaded the Old Testament; but, even though largely superseded by the teaching of Jesus, it is still available and the study of nature thus reveals spiritual truths enigmatically. We shall meet Kirby and his 'high Anglican' science again.

The great French astronomer P.S. Laplace had proposed an explanation for the solar system in which it evolved to its present stability from a mass of nebulous hot gases; and in consequence he famously told Napoleon that he had no need of the hypothesis of God. His contemporary J.B. Lamarck had suggested that animals and plants had also evolved. With these sophisms, examples of French infidelity and egotism, Kirby had little patience; and his introduction was directed against these authors and any disciples they might have. They had concentrated wholly upon second causes, and their systems rested upon a nonentity: their thinking was mere materialism, based upon the refusal to countenance metaphysical causes. Kirby was himself fascinated by the æther of Newton and Davy, an all-pervading tenuous substance responsible for many effects – but clearly physical.

Most Bridgewater authors considered that they ought to concern themselves solely with arguments for Design, so that the Bible should play only a very small part in their discussion; but Kirby refused to play by those rules. For him, God's word and works could not be separated. He noted that, in contrast to Galileo's time,[29] 'there is no danger of persecution on account of heterodoxy either in religion or philosophy. In fact the tide seems to have turned the other way, and a clamour is sometimes raised against persons who consult the revealed word of God on points connected with philosophy and science.'

With copious appeals to Francis Bacon, but also with warnings against completely separating science and religion as Bacon seemed to advise, he argued for diligence in reading, and indeed almost decoding, the scriptures. God had been

> pleased to conceal many both spiritual and physical truths under a veil of figures and allegory, because the prejudices, ignorance, and grossness of the bulk of the people could not bear them, but they were written for the instruction and admonition of those in every age whose minds are liberated from the misrule of prejudice, and less darkened by the clouds of ignorance: but still it requires, and always will require, much study and comparison of one part of Scripture with another, to discover the meaning of many of those passages of Scripture which relate to physical objects.

The reader must hunt out the meaning of a passage, for popular language does not characterise the Bible; and the evidence may be indirect and circumstantial. One great object was certainly to stop mankind from worshipping the powers of nature, rather than the true God who worked through them. The seven-branched candlestick in the Jewish Temple was thus 'a kind of planetarium' as well as having a spiritual meaning; and the cherubim surely stood for the powers of nature through which God ran the world: winds, gravity, radiation and so on. Without God they could do nothing. Kirby by using the Bible believed that he could get behind such second causes, finding intelligent design and a single First Cause.

Kirby's treatise – then and later – seemed eccentric and remote from the intellectual world in which most were moving in the 1830s. By then, other high-churchmen in England were being agitated by John Keble and J.H. Newman[30] in the 'Oxford Movement' to consider much more doctrinal questions, and were not much concerned with cherubim (or science); while liberals wanted to maintain Bacon's separation of Biblical study and science.

Whewell on physics

Much more in tune with the times was William Whewell, a prodigious know-all who held chairs successively of Mineralogy and of Moral Philosophy at Cambridge, and went on to become Master of Trinity College there and a great pundit. He had been one of the group who in the years after the Battle of Waterloo (1815) had jerked Cambridge mathematics into modernity. Through insularity and deference to the memory of its great alumnus, Cambridge had taught the differential and integral calculus in the Newtonian manner: on the Continent, Leibniz's notation was used (our familiar dy/dx).

In the hands of Laplace and others, mathematics had reached a stage where prominent Cambridge men could not follow it. Herschel, Charles Babbage,[31] George Peacock and Whewell organised the translation of a standard French textbook, and when their turns came to be examiners announced that they would include questions on Continental mathematics, thus willy-nilly modernising the syllabus.[32] Whewell went on from strength to strength in Cambridge, and became a pioneer of the history and philosophy of science as well as playing a prominent role in many scientific societies. He was chosen to write the Bridgewater Treatise on astronomy and general physics; and his, though numbered three, was the first to come out. It was a resounding success.

Whewell, an enthusiast for German gothic architecture, was aware of the writings of Immanuel Kant, and thus that God's existence could not be logically proved: like most natural theologians he therefore made no attempt to do so, but he was more conscious of what he was doing than most of them. A leap of faith was essential in religion. Whewell would have agreed with Coleridge:[33] 'Assume the existence of God, - and then the harmony and fitness of the physical creation may be shown to correspond with and support such an assumption; - but to set about *proving* the existence of a God by such means is a mere circle, a delusion'. A believer, on the other hand, investigating God's work finds out more about Him.

For Whewell, critical of Bacon in a different way from Kirby, open-minded generalising from a random mass of facts was just not possible as a scientific method. The inquirer has to get hold of the right end of the stick, by an intuitive rather than a strictly logical leap: only then will the facts form a pattern, and the interesting and relevant ones become apparent. This became, and remains, the research scientist's normal approach.

On the other hand, Whewell was deeply sceptical about the value of pure mathematics and its deductive processes. In his treatise he dealt sceptically with Laplace, and in Cambridge he ensured that as far as possible applied mathematics dominated the curriculum. Whereas Newton had inferred the law of gravity by an inductive process, his great French successors had simply worked out the consequences implicit in the law, never raising their eyes higher. 'Deductive reasoners', wrote Whewell,[34]

> those who cultivate science of whatever kind, by means of mathematical and logical processes alone, may acquire an exaggerated feeling of the amount and value of their labours. Such employments, from the clearness of the notions involved in them, the irresistible concatenation of truths which they unfold, the subtlety which they require, and their entire success in that which they attempt, possess a peculiar fascination for the intellect. Those who pursue such studies have generally a contempt and impatience of the pretensions of all those other portions of our knowledge, where from the nature of the case, or the small progress hitherto made in their cultivation, a more vague and loose kind of reasoning seems to be adopted. Now if this feeling be carried so far as to make the reasoner suppose that these mathematical and logical processes can lead him to all knowledge and all the certainty which we need, it is certainly a delusive feeling.

This statement of the arrogance of deductive thinkers may strike a chord with those who have had dealings with some mathematicians or logicians. It is perhaps surprising to find it coming from a mathematician by training, working in a university where mathematics enjoyed the highest prestige; but Whewell's interests were very wide,[35] and his commitment to religious thinking deep.

In the dedication, to Bishop Blomfield of London (a Trinity man), Whewell wrote that while he trusted that his labour would be useful[36]

> Yet, I feel most deeply, what I would take this occasion to express, that this, and all that the speculator concerning Natural Theology can do, is utterly insufficient for the great ends of Religion; namely, for the purpose of reforming men's lives, of purifying and elevating their

characters, of preparing them for a more exalted state of being. It is the need of something fitted to do this, which gives to Religion its vast and incomparable importance.

Many Bridgewater authors felt this unease, or felt the need to cover their backs against the charge of deism. Whewell's arguments in the body of the book are concerned with the fit of the laws of nature, and the extreme improbability that all the factors in the solar system and the Earth could have been so well and accurately adjusted by chance. Small differences in any one of a great number of laws and constants would have made human life impossible. He could thus carry Paley's arguments for mechanism further, adding numerous details about the length of the year, climates, stability of the planetary orbits, optics and the æther, which by the 1830s had become the medium for lightwaves. Readers would have got a readable, brief and up-to-date picture of modern physics and astronomy – clear popularisation, within a religious framework.

Whewell then invites us to think more religiously, contemplating the sublimity of nature. The vastness of the universe does not mean that God cannot love all his creatures.[37] He could do so even if all the stars were suns to other earths with their inhabitants (a view Whewell was, in a later book, to attack).[38] Equally, we should be awed but not appalled by the vast numbers of living organisms disclosed by the microscope, which had then just been much improved. Even atheists like Laplace invoked Nature to account for the uniformities they found, and could not avoid the language of final causes – functions and purposes. For Whewell and (he believed) for all reasonable people, God acts in the world continuously through the laws He has ordained.

Whewell ends with an optimistic picture of scientific progress as new fields find their Newtons and are brought into the advanced state of astronomy or mechanics; but God will still remain incomprehensible to finite human minds. Scientific progress, for him, depended upon humility, a conviction that there is an order of things, and careful inductive testing of rational but intuitive theorising – not very different from, and certainly fully compatible with, the Christian life.

Babbage: the ninth treatise

By the time the last Bridgewater Treatise (Buckland's) came out in 1836, Whewell's book had reached its fifth edition. It was extremely well received, except by his erstwhile colleague Charles Babbage who was deeply offended by the passages about pure mathematicians. Babbage wrote an unauthorised *Ninth Bridgewater Treatise* (1837) which came out from a different publisher, John Murray, and for which of course he got none of the bequest – but his book was often bound up to match the others, and shelved with them.

It is a curious volume. He was famous for his awkward disposition and, though he was a moderniser delighted to occupy Newton's chair at Cambridge, he did not feel it necessary to give regular lectures there. He is remembered for his pioneering work on clockwork computers, and as a 'character' notably enraged by organ-grinders who knew that he would pay them to desist;[39] his autobiography is highly eccentric and great fun. He had been the over-confident campaign manager for Herschel, standing against the Duke of Sussex for the presidency of the Royal Society.[40]

Babbage believed[41] that Whewell's book would 'give support to those who maintain that the pursuits of science are in general unfavourable to religion'. Yet pure mathematics dealt with necessary truths, whereas revealed religion rested on human testimony – with empirical science and natural religion coming in between. The first and greatest support for revealed religion was thus natural religion and an increasingly mathematical science: this certainly was not what orthodox clergy would want to hear. For Babbage, the truths of Natural Religion rested on foundations far stronger than those of human testimony: they were impressed in indelible characters by almighty power on every fragment of the material world. Babbage as a layman concluded that the truths of mathematics are independent of the moral character of the man who demonstrates them; and he noted, that in his profession, he did not have to follow a party line as clergy do.

There were two major features of Babbage's arguments. One was his attack on David Hume's sceptical argument about testimony, according to which its evidence would get weaker and weaker with the passage of time. Reports of miracles from long-dead witnesses therefore carried little conviction and Bible stories could thus be dismissed. Babbage, developing ideas of Laplace about testimony and probability, vigorously contested this position. The conclusion of his argument, backed by calculations in text and appendices, is that[42]

> Provided we assume that independent witnesses can be found of whose testimony it can be stated that it is more probable that it is true than that it is false, *we can always assign a number of witnesses which will, according to Hume's argument, prove the truth of a miracle.*

Mathematical argument was thus useful for answering the claims of the atheist, and Whewell had been wrong to denigrate it and other kinds of deductive reasoning.

The whole question of miracles was bound up in Babbage's main thrust, which was based upon his work on the programmable computer:[43] it was this, and not Paley's watch, which he suggested as a model for understanding the world. Laws of nature, he believed, which seem so constant to us, may change. Imagine a computer whose output we can see, and whose programme we try to determine. The 'engine' might print out the natural numbers (1,2,3 . . .) up to some number a thousand digits long; and then a different law might have been programmed into it, so that it followed a different series for a similar period, and then changed again. If ten turns of the engine's handle (making one calculation) could be made in a minute, that would mean twenty-six million in a century; and so we are contemplating millions of centuries. A series of laws thus 'successively spring into existence, at distances almost too great for human conception'. What seem to us like breaches in laws are in fact necessary consequences of some far higher law, the programming of the universe.

This made sudden changes, catastrophes, extinctions or miracles unsurprising: they seemed unexpected and startling to us, but that was because we wrongly thought we had guessed the programme. Babbage, in his roomful of clockwork, clearly had a more interesting model of the world than Paley: but it seemed to pose a problem in that apparent interventions by God are in fact predetermined features of the programme. As with other clockwork universes, there is little point in asking God to intervene, though it may make one feel better; the system is optimised and it will inevitably crush some for the greater good of the whole. This is a small consolation for the unfortunate.

Babbage's book has strange gaps filled up with lines of dashes, and nearly half its bulk is taken up with appendices. Compared with Whewell's polished performance, it makes strange reading, but it has its electrifying moments, flashes of real novelty in what was a fairly well-tilled field, where new examples were reinforcing old arguments.

Educated piety and militant secularism

What we find in the nine Bridgewater Treatises is an uncertainty about the relationship of Natural and Revealed religion; and this is something that has not disappeared with the passage of time since they came out. In what we might think of as the heyday of natural theology, authors did not believe that they had proved God's existence in the way a pure mathematician proved a theorem: they had argued a case strong enough, they hoped, to convince a reasonable person, or even to put it beyond reasonable doubt.

To the modern eye, the volumes are full of a rather cloying piety which we do not expect to find associated with popular science. This again should not surprise us, in books coming out in the waning of the Evangelical revival and the high point of the Oxford movement: there were a lot of people very full of religion compared to times before or since. Kirby's incoming tide was to turn into Matthew Arnold's ebbing sea of faith in 'Dover Beach' thirty years on.[44]

These books, however, helped to ensure that for a generation religion and science seemed in the respectable English-speaking world to be fully compatible, and indeed to reinforce each other. There were much more literal-minded readers of the Bible than the Bridgewater authors who did assail men of science; and there were also – especially among medical students and younger doctors in London dissatisfied with the status quo – rational dissenters, atheists and radicals[45] (often picking up the latest science from France) who rejected the churches' role in education and in society.

Visitors to Paris after the war ended in 1814, and again after Napoleon's 'hundred days' and Waterloo in 1815,[46] were astonished at its modern secular institutions, just as French visitors to Britain were amazed at its modern industry. In Germany and especially in France, there was no longer the same cosy relationship between élite science and an established church: separation was widely accepted and might run very deep.

Atheism, or at least militant secularism, was a powerful force among prominent French men of science, going with republican political convictions, which remained strong under the restored monarchy. The profoundly Catholic André-Marie Ampère found that his scientific career was made difficult by his faith, though for social as much as intellectual reasons,[47] while the positivist Auguste Comte[48] lost credibility when he founded the Religion of Humanity, where the key was ritual rather than sublimity or spirituality, and his calendar[49] (a revision of that inaugurated by the Jacobins) commemorated benefactors of the human race, including Saint Paul but not Jesus – his system was to be denounced by T.H. Huxley[50] as 'Catholicism minus Christianity'.

We shall later look at pantheism, where nature is seen as a source of spirituality, but meanwhile we must turn to the science that might have seemed to be most directly in conflict with Biblical narratives – geology. John Ruskin's faith was famously threatened not only by 'Puseyite' dogmatism but also by the dreadful hammers of the geologists:[51] 'I hear the clink of them at the end of every cadence of the Bible verses'. We shall find that, even here, Buckland in his *Bridgewater Treatise* succeeded in creating a space in which liberal-minded believers could work without threat to their core faith, and even find some benefit from being unembarrassed by questions about design and purpose.

CHAPTER 5

Genesis and geology[1]

CHARLES DARWIN WAS BURIED IN 1882 in Westminster Abbey, England's national shrine.[2] For his sorrowing but triumphant disciples – notably Thomas Henry Huxley, who arranged for the funeral to be there – it was a moment to look back at how Darwin had transformed geology and biology. They saw his career through the winner's eyes as a triumph of secular thinking, 'scientific naturalism', over religious bigotry: it looked uncomplicated.

Darwin's followers could reflect that England's most prominent geologist in the 1820s, the Revd William Buckland of Oxford University, had received a medal from the Royal Society for apparently discovering firm evidence of Noah's Flood. Buckland saw the Earth's history as relatively short and filled with natural catastrophes, of which the universal Flood described in the Bible was the most recent. Then his ablest student, Charles Lyell, came up in 1830 with the idea that the processes of nature were slow, and the Earth must be in consequence hundreds of millions of years old.

Lyell urged geologists to take no notice of the Book of Genesis but to make inferences about the past solely from evidence of what they could see happening now. Darwin took a copy of Lyell's book with him on HMS *Beagle*, advised by his tutor to read it but not to believe it; but his eyes were opened. In the light of Lyell's ideas, observations and collections made in South America and the Galapagos Islands led inexorably to the idea of development by natural selection. Lyell thus filled the role of John the Baptist for the Darwinians.

This story is still found in science textbooks, but the reality was more interesting. Buckland changed his mind without losing his faith. We shall focus upon Buckland – and his place in a university and nation much alarmed by the way that the Enlightenment of Diderot and Voltaire led to the French Revolution of 1789, the Reign of Terror, world war and Napoleon's military dictatorship. We shall find that Buckland did not stick fast with his 1823 proof of Noah's Flood and ignore subsequent insights. Like most contemporary geologists, he came to terms with Lyell's ideas, in a compromise that preserved his Christian belief at the cost of some reinterpretation of the Book of Genesis.

They were believers, doing research on a sensitive topic in church-dominated institutions. Rather than fighting a great battle in an endless conflict between religion and science, we shall see British scientists or 'natural philosophers' (including Lyell) closing ranks against French-inspired materialists and evolutionists on the one hand and literal-minded 'scriptural geologists' on the other. Buckland represents the general opinion in the 1830s and 1840s much better than Lyell; and his compromise endured for a generation, until Darwin's *Origin of Species* was published in 1859.

Former worlds

In his *Organic Remains of a Former World*, published in 1804, the London surgeon James Parkinson (of 'Parkinson's Disease') chose a splendid frontispiece, vividly lighting up what is to come: Noah's Ark is beached in the background with the rainbow framing it, while in front we have the remains of ammonites and other creatures that missed the boat and will be fossilised. Genesis and geology seemed to fit perfectly together. His fellow-medic James Hutton of Edinburgh had in 1785 proposed a cyclic theory of the Earth, without beginning or end;[3] the idea was revived with more clarity by John Playfair in 1802;[4] Parkinson's picture snappily refuted the cyclic theory. Since Parkinson's three volumes came out over seven years, over that time they came to need some modification.

He heard about Georges Cuvier, who had revealed in the stone quarries of Montmartre more than one former world.[5] A series of distinct faunas and floras lay fossilised beneath Paris, a discovery which implied that Noah's Flood must have been just the latest in a series of catastrophes which had wiped out life over wide areas, if not globally. These need not be floods: Cuvier pondered on the mammoth found fresh but frozen in Siberia, and inferred that an area which could support woolly elephants had suddenly been transformed into a gigantic deep-freeze.

Cuvier, based at the great Museum of Natural History in Paris, could compare his fossil bones with those of animals alive today. Guided by the principle of correlation – that all the parts of an animal must cohere – and by analogy with living species, Cuvier boldly reconstructed extinct creatures from the incomplete skeletons that he sorted out from the quarry, ever deepening as Napoleon rebuilt Paris in imperial splendour.

It was not easy at that time to accept that some species were extinct: Thomas Jefferson hoped that Meriwether Lewis and William Clark on their expedition to the West (begun also in 1804) might encounter living mastodons like those found in fossils in Ohio. There was no direct guidance from the Bible, but if God had created this the best of all possible worlds then any change must be for the worse. If species had disappeared (unless, like the dodo, exterminated by humans exercising their God-given stewardship), then it seemed that God must have made a mistake in creating them in the first place, and put it right later.

A more optimistic gloss was put upon this problem by Cuvier's older contemporary, and rival in the museum, the evolutionist J.B. Lamarck. As environments changed, so did animals: we should not weep for mastodons, but rejoice that they had progressed into elephants. For Cuvier and most of the scientific world, this seemed absurd. Animals were superb examples of correlated parts, and every species was distinct (making Cuvier's reconstructions possible); there were no missing links, and gradual change from one kind to another was never seen. Mummified cats from Egyptian tombs were just like our cats: two thousand years, a vast tract of time, had produced no change. To most people, this seemed to show that the theory of evolution was false, to the relief of those alarmed by French secularism.

Buckland: finder of dinosaurs and Noah's Flood

William Buckland[6] was a parson's son from Devon who came up to Oxford, with its newly-reformed curriculum, in 1801. After ordination he remained in Oxford, and was elected a Fellow of Corpus Christi College and Reader in Mineralogy. In 1821 a government-funded professorship was arranged for him. No undergraduate students had to attend his classes, or got credit for it, and there were no examinations. Like Davy at the Royal Institution, he had to attract and hold an audience with no set syllabus to get through, no pupils to train. Unlike Davy, there were no women at his lectures, but he attracted senior as well as junior members of the university – seen soberly gowned in a celebrated picture of the event.[7]

Buckland acquired a reputation as a speaker, entertaining as well as instructive. Oxford professors did not usually get down on all fours, for example, to show how an extinct animal must have walked: but for Buckland, as for many of his contemporaries, professing had to be a performance art. Outside the Geological Society, and even within it, there were so few experts that a full, clear and gripping style – even with earthy jokes – was essential for getting the message across. His wife, Mary, helped him to achieve a clear prose style.

Buckland was actually the first person to identify a dinosaur: an aquatic species, which did not arouse great attention. It was the later discovery of the land dinosaur Iguanodon (named for its resemblance to the iguana) by the Sussex surgeon Gideon Mantell,[8] confirmed by Cuvier, that generated real excitement; and it led to the naming of the group by Richard Owen, later the founder of London's Natural History Museum. These dragons of primeval times fascinated Buckland; but his great reputation was founded upon his investigation of much more recent material from the Kirkdale Cavern in Yorkshire. There he found the bones of hyenas and other creatures now extinct in Britain. His interpretation, in papers to the Royal Society and in his book, *Reliquiæ Diluvianæ*[9] of 1823, was that the cave had been a hyenas' den; that in the Flood they had been drowned; and that the surviving pair from the Ark had presumably (and fortunately) headed for Africa rather than back to England.[10]

The book was a success, with a new edition in 1824. It was splendidly illustrated: Buckland was allowed to use the sumptuous copper-plates that had been engraved for his paper in the Royal Society's *Philosophical Transactions*. But he needed more pictures, and these were done by the novel method of lithography. Here a drawing is made with a waxy crayon on a suitable stone, which is then wetted before an oil-based ink is applied to it. The ink adheres to the crayon marks, but not to the wet stone, and the picture can be printed. This was much cheaper than copper engraving, and came to replace it: Buckland's book marks an important transition in the visualisation of geology,[11] because books could now be more fully illustrated and their arguments thus reinforced.

Buckland, like Cuvier, made much use of 'actualism' – making present-day observations cast light on the past, on the assumption that the same general laws and processes must apply. This 'consistency over time' was later used by Lyell to replace catastrophes in explaining geology. Among the bones of the hyenas and their prey, there were also lumps of white substance that he identified as the faeces of the hyenas. In an unpleasant experiment in a menagerie visiting Oxford (which perhaps has echoes in *Alice in Wonderland*), he fed bones to the hyena there and analysed its faeces to demonstrate that they were just like those of its antediluvian predecessors. The bones it had been scrunching were marked by its teeth just as those of the hyenas' victims in Kirkdale had been. His engravings illustrated how alike were the extinct and the living species of hyena. Clearly moreover, the hyenas had lived in Kirkdale; it could not be that a corpse or two had happened to be washed up there from elsewhere.

The destruction of the Yorkshire branch of the family could not, however, be explained by any ordinary process; and it seemed to Buckland to be contemporary with great deposits of gravel, diluvium, all over Europe. This must be the effect of the Flood, which had also scooped out broad U-shaped valleys, and tumbled great erratic boulders many miles from where they had originated. It suddenly seemed as though the latest science, done by a man working in a stuffy, conservative and clerical university, had confirmed in a general way the story of the deluge in Genesis. Davy visited Kirkdale with Buckland, and was impressed. Buckland became a scientific star.

In 1822, before the book had come out, Buckland was on the strength of his papers awarded the Copley Medal of the Royal Society. This was its highest honour, and had never before in the ninety years since it was established been conferred for 'pure geology'. Davy as President made the speech on 30 November 1822. Cuvier had first elucidated the great mystery of extinction, but Buckland had provided decisive evidence that now-exotic creatures had once existed where their remains were found. A change in climate, a cooling of the Earth, seemed to be implied; which was not surprising given its hot centre. Davy added that[12]

> it is gratifying to feel that the progress of science establishes, beyond all doubt, the great catastrophe described in the sacred history . . . and that it likewise demonstrates . . . a successive creation of living beings, of which man was the last, destined to people the earth, when its surface had assumed a state of order and beauty fitted for the improvement and activity of an intellectual and progressive being.

This was not, Davy said, a brilliant speculation after the fashion of the Enlightenment, but a geological deduction.

Sound Baconian reasonings had replaced visionary dreams of eternal cycles [Hutton] or evolution [Lamarck, Erasmus Darwin]; but just as dreamy or bad would be attempts to frame 'systems of geology out of the sacred writings, by wresting the meaning of words, and altering the senses of things'. The laws of nature, and the principles of science, were to be discovered by labour and industry, and have not been revealed to man 'who with respect to philosophy [science], has been led to exert these god-like faculties, by which reason ultimately approaches, in its results, to inspiration'. We have no right to measure divine truths by our fancies or opinions: 'they should be kept perfectly distinct'.

Turning to Buckland, Davy then presented the award, noting the light that geology casts upon the 'economy of nature', making the globe varied and habitable; and noting too its capacity to fire the imagination.[13] If we marvel at the ruins of Antiquity, how much more must we wonder at

> those grand monuments of nature, which mark the revolutions of the globe: continents broken into islands; one land produced, another destroyed; the bottom of the ocean become a fertile soil; whole races of animals extinct, and the bones and exuviae of one class covered with the remains of another; and upon these graves of past generations, the marble or rocky tombs, as it were, of a former animated world, new generations rising, and order and harmony established, and a system of life and beauty produced, as it were, out of chaos and death; proving the infinite power, wisdom, and goodness of the Great cause of all being!

Banks had cold-shouldered geology, but now it was welcomed into the scientific fold. Buckland's readers were warned that the methods of science and of religious interpretation were quite different, and must not be confused; but invited to rejoice that, as Bacon had foretold, they were compatible. Buckland's phrase 'creative wisdom', for example, struck a chord with John Ruskin,[14] to whom Buckland had taught science and to whom he introduced Darwin.

Religion and science

The tension between Dissenters and Anglicans, Chapel and Church, is crucial to understanding British history in the eighteenth and nineteenth century as the country creaked from oligarchy towards democracy and pluralism. The Reformation had left Great Britain a firmly Protestant island,[15] but with a Presbyterian established Church of Scotland, and an Episcopalian Church of England. Roman Catholics were feared and disliked by almost everybody. There had been religious toleration since 1689, but Dissenters were second-class citizens and some, like Priestley, bitterly resented it. Their numbers were swelled in the later eighteenth century when the Methodists separated from the Church of England. Barred from the ancient universities, and from government posts, Dissenters set up their own academies, and took to business and industry.

Men of science did not rally to Priestley, the radical Unitarian, after his house was sacked in the Birmingham riots, and in 1794 he departed to exile in the USA. The cause of Church and King, orthodox religion and constitutional monarchy, seemed the right one in the face of French success in exporting the revolution by force as the Grand Army marched into Italy, Germany, Spain and Russia. Most men of science in Britain had always been religious, though not always orthodox, and there were very few avowed atheists (though not everybody seemed to fear the Last Judgment), but it became imperative to show that useful scientific knowledge also supported the social and religious system.

Combining the 'book of nature' with the scriptures had been Francis Bacon's recommendation for wisdom and humility at the outset of modern science in the seventeenth century. Science as a route to God had ever since seemed something upon which Dissenters and Anglicans could agree, and even work together. Paley's *Natural Theology* became the classic statement of this notion.[16] Studying God through His works was a long-standing enterprise, but natural religion based entirely upon this 'book of nature' was the source of the French Revolutionaries' Festival of the Supreme Being, intended to replace the traditional festivals of the Church – and thus anathema to good Christians.

Even so, unbelieving France (meaning Paris) was around 1800 the world's centre of excellence in science.[17] From astronomy and botany through to zoology, the often-irreligious leading men of science were there: associated particularly with the Academy of Sciences, suppressed as élitist after the Revolution but soon reopened. Cuvier and Lamarck's museum was never shut down; its lectures and gardens were open to the public, there was a zoo nearby[18] and Britons contemplated it with envy. They were aware that London science was by comparison under-funded and provincial.

The learned world

The English universities, Oxford and Cambridge, were primarily engaged in training clergy for the Church of England; they also provided a finishing school for young gentlemen. Paley's books became recommended reading for freshmen at Cambridge, though not at Oxford where utilitarianism was disliked. The emphasis in both places was on a liberal, rather than a useful, education. About 1800, both universities modernised their courses: Oxford came to specialise in classics and philosophy, Cambridge in mathematics. Able students, competing for honours, had little time for anything else. Professors were appointed to give optional lectures in other disciplines,[19] but their role in the university was marginal compared to that of heads of colleges or the tutors who taught classics or mathematics.

Scotland was different. The little city of Aberdeen had two universities, like all England; but internationally it was the universities of Edinburgh[20] and Glasgow that were significant, for their medical training and their lack of religious tests. Nevertheless it was in educationally-conservative England that the industrial revolution had begun (often associated with Dissenters), ensuring that Britain's economic power would defeat the French.

In London, which was increasingly attracting talent from Scotland as well as the English provinces, the Royal Institution was set up in 1799 to promote applied science. Its great success was in finding Humphry Davy,[21] a brilliant lecturer and researcher whose electrochemistry won him a prize from the academy in Paris. He went to collect it, despite the war, putting British science on the map in the hostile capital city. On his return in 1815, he invented a safety lamp for coal miners, proving that an instrument devised in a London laboratory by an eminent scientist would work in the subterranean world of heavy industry. As the gentleman with the lamp, Davy was the unstoppable candidate for president of the Royal Society when the post became vacant in 1820.

Meanwhile, canal engineers and mining managers – 'viewers' – were building up an empirical science of geology, distancing themselves from the big picture and systems like Hutton's. In 1807 the Geological Society was set up in London. Davy, who had grown up in the tin-mining county of Cornwall, was enthusiastic; but Joseph Banks – who, as president, was piloting the Royal Society, Britain's Academy, through the difficult years of revolution and war[22] – believed that if science were fragmented into specialisms then it would be lost. All he would allow was specialised groups within the Royal Society. He was Davy's patron, and made Davy resign from the infant society.

It nevertheless flourished, and developed a controversial style in its meetings In the Royal Society, papers were read to an audience sitting in rows facing the formally-dressed president enthroned behind a handsome mace, and discussion was not permitted; there is a splendid portrait of Banks formidably presiding.[23] At the Geological Society,[24] seats faced inwards, as in Parliament, with the most prominent members occupying the front benches, and debate was the order of the day.

Geology became one of the most exciting and prominent sciences of the nineteenth century. Britons began here, as in chemistry, to fight back against French superiority. In a project that began with the engineer William Smith, geological strata were defined (and patriotically named Cambrian, Silurian, Devonian and so on) from their characteristic fossils. Was there anything in the Bible to account for these strata and fossils? If not, did that matter? Perhaps the Bible wasn't meant to explain everything?

More than one way to read the Bible

Christians had been accustomed to read the Bible in a variety of ways.[25] There were messages for the perceptive and reverent reader at different levels; indeed, the text was used in what we might call a post-modern fashion, where meaning is imposed by reader and context. The story of Noah tells us about God's wrath at sin, and his saving power, as well as about the Flood; and refers forward to the Church, an ark with room for everyone floating upon the world's tempestuous sea and safely conveying us to the heavenly Jerusalem. This is how the story is used when it is read in the liturgy along with passages from an epistle and a gospel.

While our ancestors took it for granted that the Bible was the Word of God, and literally true, they read or heard much of it as they would a great poem or play: to be mulled over and understood at more than one level. The problem was whether such understandings would all collapse if the story were not literally true. Certainly, stories of miracles were by the nineteenth century becoming an embarrassment to sophisticated believers; and it was a matter of sometimes-agonised conscience (as for Philip Henry Gosse)[26] whether to accept scientific inferences or the apparently plain sense of scripture.

Partly because Galileo had been punished by the Pope and Inquisition, all good Protestants knew that he must have been right; and nobody in Britain in 1800 seriously believed that the Earth was flat and the Sun went round it, although that is how some Biblical passages might well be read. Clergy and laity accepted that 'the Bible was to tell us how to go to Heaven, not how the Heavens go'. Some passages in the scriptures are much more important for the drama of human salvation than others. Probably few people believed in the story of Balaam's ass, which saw the angel blocking its master's way and spoke to him: the story could be understood simply to mean that sometimes a dumb animal may be more perceptive than its sophisticated but asinine owner.

If in Buckland's time the centre of things in science was France, in theology it was Germany; Britain was provincial. There was anxiety about what philosophers, liberal Protestants and biblical critics – investigating the Bible like any other ancient book – might be up to there, but few spoke German. Buckland's friend and colleague, the theologian E.B. Pusey, visited Germany to find out[27] and disliked what he saw.

Texts about astronomy are scattered about the Bible, and do not seem crucial to it: Galileo saw no theological limitations on his study of the heavens. The creation story (or stories, because chapters 1 and 2 of Genesis are separate) was rather different. It comes at the very beginning but it is also echoed near the end, in St John's Gospel. God saw that His creation was good; but through the disobedience of Adam and Eve the world was marred by sin and death. Only with the incarnation, death and resurrection of Jesus, the second Adam, could mankind be saved. This whole scheme, evident enough in the Bible and set out magnificently in John Milton's epic *Paradise Lost* (1667), was of immense importance to believers.

Milton's contemporary James Ussher, Anglican Archbishop of Armagh, had calculated the date of the creation from the ages of the various patriarchs recorded in the Old Testament and had come up with the figure of 4004 BC. This date (not very different from less authoritative ones that had been around for centuries, like 4600 BC in the York Mystery Play, *The Harrowing of Hell*) found its way into some editions of the Bible, but never became a part of church dogma. Nevertheless, many believers took it as plain exposition of the sacred text, and a serious matter. On a nineteenth-century Masonic tombstone in St Oswald's churchyard in Durham, the dates 'of masonry' have been computed by adding 4004 to the AD figures, making masonry as old as the creation.

Dissenters and Tractarians

The late eighteenth and early nineteenth centuries were a time of religious revival and change. They also marked the coming of avowed unbelief; but most geological debate was between believers. The old Dissenters were the descendants of the Puritans, some of whom had sailed to America; they were Calvinists, and might be Presbyterians or the more loosely-organized Congregationalists. By the late eighteenth century, many reacted against the strict Calvinist doctrine that a minority were in God's foreknowledge elected for salvation and a majority were marked for damnation. Like Priestley, they became Unitarians, that featherbed for Darwins and Wedgwoods.[28] Thoroughly liberal, sharing the values of the Enlightenment, they became patrons of science and secular learning in Britain and the new USA; but Unitarianism never became a faith for the masses.

To the rural and industrial poor, Wesley's enthusiastic preaching (with its promise that salvation was open to all who claimed it)[29] was welcome; and the Methodists grew rapidly, becoming the new Dissenters. Within the Anglican and Presbyterian churches, those who stayed were often deeply affected by Wesley, and by 1800 there was a major evangelical revival. Prominent evangelicals living in what was then an attractive suburb of London were called 'The Clapham Sect'; they included William Wilberforce. Interested in social issues (especially the abolition of slavery), they were by no means averse to science. In Scotland, Thomas Chalmers the Bridgewater author was an evangelical minister. They could agree to co-operate (to the horror of high churchmen) across denominational boundaries in campaigning on social issues, and in promoting the translation and circulation of the Bible.

After Waterloo in 1815 came peace and economic crisis. Political change in Britain, shelved for the duration of the war; now became urgent. Votes were demanded for all men who owned property or practised a profession, and thus had a stake in the country. Religious toleration no longer seemed enough: Protestant Dissenters in Great Britain, and the Roman Catholic majority in Ireland, demanded full civil rights. During the 1820s, after various crises, even Catholic Emancipation was carried through Parliament: Davy, who loved Ireland, Italy and Slovenia, was one of the few men of science to be really delighted about that. Most Scots and Englishmen were reminded of the Spanish Armada, the Inquisition and Galileo, and felt uneasy: 'No popery' remained a rousing cry for many more years.

Then in 1832 came the Reform Bill, essentially giving the vote to the middle classes and abolishing 'rotten boroughs' where a handful of electors had chosen a Member of Parliament. The king was the Supreme Governor under God of the Church of England, and Parliament legislated for it. A reformed Parliament might reform it, take away its endowment and open Oxford and Cambridge to all comers. The prospect of this national apostasy horrified John Keble, one of the bright stars of Oxford in the 1820s, famous as a poet[30] and hymn writer (his *Christian Year* far outsold William Wordsworth's poetry) and now a country vicar. Invited in 1833 to preach before the judges in Oxford he denounced it, striking a chord among conservatives and churchmen in the face of reformers and Dissenters. Pusey and John Henry Newman, one of the century's great intellectuals, joined Keble in what became the Oxford Movement,[31] making the evangelicals among whom they had grown up their prime target, as Frank Turner shows.[32]

Because they began with the publication of short tracts, they were at first named 'Tractarians'. Not only were they hostile to the pretensions of dissenters, but they also sought to recall the Church of England to its Catholic past. They saw it as the true heir of the early church. The reformers of the sixteenth century had swept away medieval accretions, but in the process they had undervalued ritual and liturgy, and the vocation of the priest. Penitence was appropriate. Order and dignity should be brought into worship, and church buildings should be restored to something like their medieval condition. A wave of religious enthusiasm spread through Oxford, and its ripples reached Cambridge. For almost a decade, students flocked to hear Newman's sermons, while debates about doctrine and worship drove out any concern with secular subjects such as the empirical sciences.

Did God intend us to study geology?

Meanwhile, the question of the Earth's creation – and what to believe – had burst on to the public stage in a series of books that brought geology into prominence and changed people's ideas about the history of the Earth. William Buckland's *Reliquiæ Diluvianæ* (1823) was followed, in the early 1830s, by

Humphry Davy's posthumous *Consolations in Travel* (1830), Charles Lyell's *Principles of Geology* (1830–33) and William Buckland's *Geology and Mineralogy Considered with Reference to Natural Theology* (1836), the last of the Bridgewater Treatises to be published. As the churches more and more concentrated on religion and turned their backs on science, so the scientists felt increasingly free to define and explore their subjects unhindered.

Declaring geology a fully respectable science, akin to astronomy in its sublimity with endless vistas of time[33] rather than of space, was not one more round in a contest between science and religion. Instead, geologists (many of them clergymen) were given the backing they needed to debate the age of the Earth and the nature of fossils, and eventually to sideline their literal-minded co-religionists. Buckland was on dangerous territory, as anybody is who allies the latest science with close Biblical interpretation. For him, authority was to be found in quarries, outcrops and caves. If it confirmed Biblical accounts, well and good – if not, that was a sign that literal interpretation was at fault. Others – and the Revd William Cockburn, a Cambridge scholar who became Dean of York was most prominent – denounced the whole attempt to test the Bible against empirical evidence, and saw Buckland's work as mere delusive hypothesis in which the scriptures were exposed to criticism.

In 1838, the Revd George Young, who had published a survey of the Yorkshire coast, tried to read his paper 'Scriptural geology' to the BAAS, meeting in Newcastle; but was restricted to the abstract. He published the full text as a little book,[34] arguing that a few weeks after the Flood would have been enough to consolidate the strata. Another writer from the north-east of England, Thomas Cooper, published his lectures on the *Mosaic Record of the Creation* in 1878,[35] from a surprisingly respectable publisher given its date – we should not suppose that everyone went along with Buckland.

In Ireland, the prominent lawyer Dominick McCausland argued in *Sermons in Stones* (1857) that modern geology (with a range of references to Buckland, Lyell and Roderick Murchison) was fully compatible with the Book of Genesis taken literally: and his book went through thirteen editions.[36] But in America, Edward Hitchcock, President of Amherst College, in his *Religion of Geology* defended the infallible basis of scripture while at the same time denouncing scriptural geologists as factious, and deplored Kirby's close adhesion to the literal text; ending up with a catastrophism not unlike that in Buckland's later writings.[37]

Buckland's activity was looked at askance, not just by literalists but also by those who feared that faith and morals might be damaged by unseemly debates about biblical truth. He had dedicated his 1823 book[38] to the liberal and intellectual Bishop of Durham, Shute Barrington, with thanks for his 'indulgent notice of my endeavours to call the attention of the University to the subject of geology' and he referred to 'the inestimable and most judicious work' of John Bird Sumner,[39] an evangelical exponent of the Design argument who later became Archbishop of Canterbury. Buckland's conclusion was that there had been a recent and universal inundation, whose cause was unknown, and that this Deluge happened at about the time recorded in Genesis, before humans reached Europe.

Yet the last paragraph of his main text alludes to older rocks, shattered and disturbed by violent convulsions and eroded long before the Flood. To those who came to call themselves scriptural geologists, Buckland was playing fast and loose with Genesis, accommodating Noah but pushing the date of creation back by millions of years. Liberal-minded Christians might at this point argue that the Bible was to tell us how to go to heaven, and not about mastodons; but for the literal-minded this would not do. The Royal Society, and then in the 1830s the British Association for the Advancement of Science, and new Church institutions like King's College London and the University of Durham, provided important room to manoeuvre for the geologists like Buckland, and his opposite number at Cambridge the Revd Professor Adam Sedgwick with whom Darwin had geologised in Wales.

Lyell's tracts of time

Charles Lyell (1797–1875) was a pupil of Buckland – who, like the best professors, changed his views when they were tellingly criticised by an able student. As the layer of diluvial gravel (found apparently all over the world) was more closely investigated, so it seemed that it was of various ages. Instead of indicating a universal deluge, it fitted only with a scheme of more local catastrophes. Nor could Buckland account for the catastrophes that transformed one order of things into another: God seemed necessary to close that gap. Buckland knew about rain, snow and ice, about rivers, about the ocean and its icebergs; but he found it difficult to practise 'actualism' – the study of natural changes as they were happening – since Britain was geologically old and stable.

On the other hand, Sir William Hamilton[40] as a British diplomat in Naples had made extensive studies of Vesuvius; Cook's expeditions had observed volcanoes, as had Banks in Iceland. Even so, the convulsions of nature, in eruptions, earthquakes and tidal waves, while appealing to Romantic ideas of the sublime, seemed remote. Hutton and Playfair had invoked the power of fire as an antagonist to that of water: these agents were named after ancient Gods, 'vulcanism' and 'neptunism'. Then Lyell visited Italy, a geologically-recent country.

In the ancient rocks of Britain, the fossils are very different from the creatures now living; in Italy this is not the case. Lyell found that he could make sense of the relatively-new 'Tertiary' rocks, which he divided into eocene (the oldest), miocene and pleistocene, by seeing what proportion of the fossils belonged to still-living groups. Statistics provided the key to dating and the boundaries between the epochs were not natural frontiers, marked by mass extinctions, but were a matter of convenience (like our division of history into centuries). Deep time was a continuum, and nature made no jumps.

Lyell came to believe that Buckland had had to invoke vast and inexplicable catastrophes, wrongly bringing divine action into science, simply because he had not allowed long enough for the ordinary processes of nature to do the job. The frontispiece of Lyell's *Principles of Geology* shows the Temple of Serapis at Pozzuoli.[41] Its columns are marked about one-third of the way up by sea organisms. Clearly, it was not built under water; but, in its history since ancient Roman times, it has been submerged and then elevated again – and this has been so gentle a process that the columns are still standing. They are contemplated by a seated figure musing in the foreground.

Like Parkinson's *Organic Remains of a Former World*, Lyell's book was in three volumes and came out over several years (1830–33). It set out a radical separation of geology and Genesis. The book was an attempt to explain the past exclusively in terms of processes now operating, as the essential basis for geological science. This involved drawing upon immense tracts of past time. The relative dating of strata which had so far been done left absolute durations open: nobody could say, for example, how long the Devonian period had lasted.

Lyell used evidence from the rate at which coasts had been eroded and river deltas deposited in historical times: and extrapolated from hundreds or thousands of years to the hundreds of millions he needed. In his uniform world, there had been little serious change; and in particular he disliked the idea of progress, which he associated with both religious apologetics and Lamarckian evolution. For the former, he had no time; and he used a whole chapter to refute the latter. Progress through the strata from fish to reptiles to mammals was probably illusory; and as more geological research was done, so a more uniform picture would become evident.

The geological record was known only from western Europe and the eastern parts of North America, the process by which creatures became fossilised was very chancy, and thus existing data were very imperfect. Lyell was trained as a lawyer, and his presentation of his case was very skilful; but it seemed to most contemporaries too extreme to be wholly plausible. The young Charles Darwin, setting off on HMS *Beagle* in 1832,[42] was to be Lyell's most important convert and disciple, using his framework to bring the idea of evolution successfully (though belatedly) into biology.

Davy's Consolations

As a statement of the views he criticized, Lyell used a book just put on sale by his own publisher, John Murray. This was Davy's posthumous *Consolations in Travel* (1830). Following a stroke, the prematurely-aged Davy had returned to his beloved Julian Alps (in Slovenia), wintering in warmer Italy; and, contemplating imminent death, he composed a series of dialogues about science and life.[43] The first dialogue presented a dream of progressive history, where science and technology had been the key to improvement and a few gifted individuals responsible for it; this picture was contested by another speaker, as unbiblical.

A book of dialogues (in the tradition of Plato, Galileo, Boyle and Hume) allows an author to advance various opinions and explore uncertainties – but we can take it that, usually at least, the progressive view was Davy's own. In other dialogues, Davy drew attention to petrifying springs in Italy, and explored their analogies with the agents that had produced the fossils in the secondary rocks. Davy had also spent some months in Naples while he attempted to unroll manuscripts found in Herculaneum. While there he visited Vesuvius, which led him to see both water and fire as sometimes destroying, sometimes creating, mountains and islands.

A mysterious stranger, the Unknown, takes up the exposition of geology: as the Earth cooled, the primary rocks (without fossils) crystallised and the water contracted. The shellfish and coral 'insects' of the first creation began their labours, and islands appeared in the warm sea. They became covered by vegetation, like palm trees, fitted to bear high temperatures; and shellfish were joined by other fishes, while the cooling fluids of the Earth deposited more solid matter to mix with their remains. Species of egg-laying reptiles were created next; but[44]

> in this state of things there was no order of events similar to the present;- the crust of the globe was exceedingly slender, and the source of fire a small distance from the surface. In consequence of contraction in one part of the mass, cavities were opened, which caused the entrance of water, and immense volcanic explosions took place . . . When these revolutions became less frequent; and the globe became still more cooled, more perfect animals became its inhabitants.

Since then there had been a great upheaval, which was the cause of those remains usually called diluvial and probably 'connected with the elevation of a new continent in the southern hemisphere by volcanic fire'. The creation of man had taken place only when things had become more permanent.

Change without evolution

There followed a discussion of how this scheme fitted in with Genesis, and whether the creation of the world might (with advantage) have been left out of this view altogether.[45] The Unknown, with reference to Hutton and Playfair, declared that he had 'no objection to a *refined plutonic view*': but could not accept complete uniformity. There was progress through the strata, each of which contained 'remains of peculiar and mostly now unknown species of vegetables and animals', and it was impossible to study these fossils 'without being convinced, that the beings, whose organs they formed, belonged to an order of things entirely different from the present'. He made use of the term 'diluvian' merely to signify loose and water-worn strata, not at all consolidated and deposited from an inundation of water. Where they were found, mankind did not exist, so they should not be identified with the deluge described in the Bible. There must have been a series of destructions and creations, a gradual approach to the present scheme of things, in preparation for mankind.

What would be absurd would be the atheistic evolutionary view 'that the fish has in millions of generations ripened into the quadruped, and the quadruped into the man'. Equally absurd would be the opposite idea, based on a literal understanding of the Bible, that God had created in an instant everything that geologists had since found and 'that the secondary strata were *created*, filled with remains as it were of animal life to confound the speculations of our geological reasoners'. On this view, fossils were in effect fakes, as (Phillip) Henry Gosse FRS of the Brethren church later seemed to imply in 1857 in his book *Omphalos*.[46] He suggested that just as Adam and Eve would have had navels falsely indicating that they had had mothers, so the trees in the Garden of Eden would have had rings suggesting unreal age and the rocks would have contained fossils of creatures that had never existed.

Why Lyell picked upon Davy is not clear; but perhaps it was because he was very eminent, he was dead and could not answer back, and his was a lucid statement of the idea of geological progress. Lyell criticised Davy's progressionism, but accepted his view that mankind was a recent creation. The bishops and other churchmen who governed King's College London had no objection to millions of years of pre-human Earth history, provided it was not combined with evolution, and Lyell was appointed Professor of Geology in the new institution; though this Anglican college later barred Justus Liebig the chemist from consideration as a Lutheran.[47]

William Buckland also bought Davy's book and marked some passages. Davy had convinced Murchison, a military man unemployed after Waterloo, that he could enjoy both science and field sports and as a result he became one of the great geologists. Drawings in Croatia and Slovenia by Murchison's wife were engraved for later editions of *Consolations*,[48] which sold steadily and was translated into Spanish, German, French and Swedish. Davy, Murchison, Buckland and Sedgwick better represent the position most geologists took in the 1830s than Lyell does.

Buckland on geology

Then Buckland was invited to write a Bridgewater Treatise: his (published in 1836) was the last, although numbered six out of the eight. It was recognised as one of the best. It has a second volume of plates; it was said that he spent his £1,000 on these illustrations, which are magnificent. The series was sometimes unfairly called the Bilgewater Treatises; Murchison, writing to Lyell, called this book the Bridge over the Water, because in a long footnote[49] Buckland repudiated his previous identification of Noah's Flood with the water that had drowned his hyenas:

> Discoveries which have been made, since the publication of this work, show that many of the animals therein described, existed during more than one geological period preceding the catastrophe by which they were extirpated.

What Buckland now envisaged was more like Davy's world: progressive change in an Earth gradually got ready for human occupation. Extinct creatures did not die out because they were poorly designed; God loved them, and had admirably contrived them for their time and place. Buckland's 1823 book went back far enough to accommodate Noah's Flood, and not much farther; by contrast, his 1836 treatise involved an immensely long past.

Buckland's first illustration is a hand-coloured lithograph, which folds out to well over a yard long (more than one metre), showing the Earth's long history from left to right. The characteristic animals and plants of each epoch are delicately drawn above. On the extreme right, we have a volcano and a dodo (connections to *Alice* again; there was a stuffed one in Oxford). The effect is surprisingly modern and the progress from the oldest marine invertebrates up to familiar mammals (mankind is tactfully omitted) is evident. Buckland was not prepared to go all the way with Lyell, but his actualism was reinforced by his student's writing even though various convulsions remained part of his scheme.

The Bridgewater Treatises were not supposed to be directly concerned with revealed religion, but Buckland as a parson–geologist had to indicate where he stood with respect to Genesis. It is not really very difficult to allow for millions of years, while respecting the literal text: God's time is not ours, and each of the six days might be a thousand ages to us. That was not Buckland's solution. He believed that science could help us interpret scripture rightly and he went along with other scholars, for whom there is a break between the first verse of Genesis – 'In the beginning God created the heaven and the earth' – and the second: 'And the earth was without form and void . . .'. That first verse is about God and the created order, a theological statement; and the phrase 'in the beginning' covers all the period up to the creation of mankind. Knowing about iguanodons or mastodons is not essential to salvation.

The story then leads into the account of how the modern order of things was set up: 'the earth . . . without form and void' had gradually changed over aeons, long before humans existed. The whole study of fossils – and 'palaeontology' was by the 1830s the most exciting part of geology – was concerned not simply with times that were pre-historic, but were in a sense pre-biblical. We might suppose that this view was some eccentricity of Buckland's, but it wasn't. His friend Pusey endorsed his account with a substantial note,[50] going on for four pages, and full of curious learning. In 1847 the Scottish churchman George Wight in his *Mosaic Creation Viewed in the Light of Modern Geology* also favoured Buckland's interpretation; it was not some high-church Anglican oddity.

Seeing Design everywhere, Buckland was particularly delighted with the discovery of coprolites, which appeared to be the fossil faeces of dinosaurs. He could extend his earlier studies of the bowel movements of hyenas back millions of years; and show how efficient and well-contrived the digestive systems of early reptiles had been. Extinct organisms had died out not because of any design faults, but because the circumstances to which they were appropriate had changed with slow global cooling: thus the vegetation of the Secondary period was more tropical in character than that of the Tertiary. God had certainly not blundered His way towards the present state of things by trial and error, burying His mistakes like a bad doctor.

Buckland was also delighted by the discoveries of footprints and tracks of extinct creatures, and reproduced some of them full-size on folding plates. He noted undeviating unity of design in the animal and plant kingdoms; remarking how extinct animals often provide links between existing species. He followed the improvements in eyes from the trilobites onwards, fascinated by correlations and developments of a theme. Tree-rings, he noted, could be used as a chronometer. Design was even evident, order amongst apparent confusion, in the turbulent disturbing forces that diversify the land, make minerals available and supply water through springs.

As one might expect of someone living through the early Industrial Revolution, Buckland (like Parkinson) took a great interest in coal. He remarked upon the importance of being economical in our use of it; and saw the existence of beds of coal and ironstone (often adjacent to each other in Britain) as an indication of divine providence and foresight. Not only had the primitive forests been fossilised as coal, but these deposits had then been upheaved so as to be accessible to us. And the fossil record seemed to Buckland to show that while there had been long periods of slow and gradual accumulations, there had also been times when large numbers of creatures had died suddenly, and been entombed before they began to decay:[51] thus octopuses 'must have died *suddenly*, and been *quickly* buried in the sediment that formed the strata, in which their petrified ink and ink-bags are thus preserved'. The evidence did not square with Lyell's unyielding actualism.

Readers of Buckland's Bridgewater Treatise would thus have got an astonishingly vivid account of the primeval world. They would appreciate what extinct animals were like and how they lived, and how new discoveries (he reported on those of the young Charles Darwin from South America) were all the time modifying the provisional conclusions of men of science. Nevertheless, some conclusions would stand. The message that God's loving design was as clearly visible in extinct as in living creatures extended Paley's reasoning back to an indefinitely-distant past.

And, just as Cuvier, Davy and Lyell had rejected evolution, so did Buckland. He needed acts of special creation for the various species rather than any kind of development. There were more gaps in his system for God to fill than were left by Lyell, who nevertheless had needed some acts of special creation, notably for mankind. Indeed, Lyell, despite being in on Darwin's secret, remained unconvinced of evolution until well after the *Origin of Species* was published in 1859.[52]

No Ice Age in the Bible

Cuvier's mammoth quick-frozen in Siberia seemed evidence for a sudden and terrifying cataclysm; Lyell found a more naturalistic explanation,[53] drawing upon stories from Spitzbergen of crevasses and avalanches, where animals might fall in or be overwhelmed and frozen – but there was more to ice than that. Among the unconvinced was the Swiss geologist Louis Agassiz (1807–1873) who had worked with Cuvier and became the great authority on fish. He also studied the glaciers of the Alps.[54] After the cold winters of the seventeenth century, the climate had begun to warm up. By 1800 the glaciers had started to retreat; and, where they had been, Agassiz found moraine, erratic boulders and polished rocks.

He camped on glaciers to study how they moved and described his findings in a book, *Etudes sur les Glaciers* (1840). It has superb lithographed illustrations of Swiss glaciers, with transparent overlays that contain the explanations, superimposed on the facts. In 1837–8 Agassiz had suggested that the effects supposed to be due to the Deluge were actually produced by glaciers that had spread over much of Europe and North America. Buckland visited Switzerland in 1838, and was convinced; in June 1840, under his presidency, a paper by Agassiz was read at the Geological Society. Agassiz came to Britain, and addressed the British Association in Glasgow that September. After the meeting, he toured Scotland with Buckland, and everywhere they found the familiar evidence of glacial action.

Buckland converted his former pupil Lyell, who was at first reluctant; and in the face of opposition, notably from Murchison (who attributed the effects to icebergs grating along in a shallow sea), in great debates during the winter, the idea of an Ice Age gradually gained ground.[55] Its cause was unknown, but it could better explain the evidence than Buckland's Flood had done. Agassiz became one of the first prominent European men of science to be tempted down the 'brain drain' to the USA, when he accepted a post at Harvard, making studies of American fauna and emerging later as one of Darwin's prominent opponents.

The slow advance and retreat of glaciers was very different from the sudden freeze-up invoked by Cuvier to explain his mammoths. While climatic changes clearly can be catastrophic, they did not require special interventions by God; and Agassiz had reasoned in the manner prescribed by Lyell, extrapolating from processes happening in the present day. Accepting an Ice Age did mean that the uniform cooling of the Earth envisaged by Davy and others was no longer an option; but it left open the question of whether there was progress over time, with later animals and plants being further up the scale.

There was no mention of the Ice Age in the Bible, so the link between Genesis and geology, which had impressed Buckland and even Davy in the 1820s, had gone by the 1840s. What was left was the belief that in a general way, the beginning and the days of creation (rightly interpreted) were compatible with the emergence of the present state of things, and especially the recent appearance on the scene of men and women: beings with moral sense, language and intellect denied to mere animals.

Huxley: Darwin's bulldog

Deep concern with the material world seemed to run counter to deference, habits of obedience and real, spiritual religion. Newman was no biblical literalist – indeed, he provided much ammunition for sceptics – for he realised the dangers of ignorant readings of the scriptures and appealed to the tradition of the early church as guide. The focus on religion, especially a traditional and medieval 'Catholic' version, made science marginal. The bubble burst when in 1845 Newman converted[56] to Roman Catholicism. About the same time, the modernising of Oxford on the German model, into a secularised research university offering degrees in many disciplines, began – but it was to take a generation.[57]

Universities in Europe look for recognition by the government, a charter that allows them to grant degrees and lays down how they shall be governed. But in 1828 a self-styled London University was set up, soon denounced as a godless institution because it was secular. In fact, its backers and early students were usually not godless, but Dissenters of one kind or another or Jews.[58] Only when it later (as 'University College') federated with the Anglican 'King's College' into the University of London did it get its charter. London was by then attracting medical students in large numbers and the new university thus took a scientific direction.

The evolutionary *Vestiges of Creation* (1844) created a sensation; and among medical students, like the young T.H. Huxley[59] who coined the term 'agnostic' and became Darwin's bulldog, there grew up a contempt for organised religion – impotent in dealing with the social problems of London's slums – and for effete Oxford or Cambridge clergyman–professors. Huxley belonged to the generation which would declare a war between science and religion, and indeed skirmish with William Wilberforce's son Samuel, Bishop of Oxford, at the British Association in 1860.

Huxley's frontispiece for his book, *Man's Place in Nature*[60] (1863) has become an evolutionary icon. It shows from left to right the skeletons of a gibbon, an orang-utan, a chimpanzee, a gorilla and a man, forming a kind of procession. It was meant to demonstrate how we and the modern apes form a natural group, like horses, donkeys and zebras; but to all who had read Davy or Buckland and absorbed the idea of progress, it looked like our family history and was duly shocking. For Darwinians, we are not descended from the orang-utan or any other living species of ape; but we share a common ancestor with them in the not-so-very-remote past. Huxley's frontispiece got across his message powerfully – if ambiguously – as Parkinson's, Lyell's and Buckland's had done earlier.

Huxley's way with words and pictures, his love of a good fight, his earnest advocacy for science, his great abilities and his charm propelled him into the presidency of the British Association and then of the Royal Society. By the time he died the ideas that Cambridge undergraduates should study Paley, that first-rate universities should be denominational, or that theology was in any way Queen of the Sciences had all gone.

For Huxley and his contemporaries, all this was (as Keble, Pusey and Newman had foreseen and feared) part of the struggle that had begun in the 1820s to break the stranglehold of Old Corruption and privilege on government and education. Although it is curious that Britain entered the twentieth century under the administration of a gloomy, churchy aristocrat, Lord Salisbury,[61] nonetheless science was by then fully emancipated from such connections. In the USA, state and church had always been separate; and by 1900 the research ethos, and a weakening of denominational ties, had become a feature of the great universities there.

It is tempting to look back at Buckland through Huxley's eyes and to see conflicts between the great abstractions, religion and science, going on in and around him. But, when instead we come to him forwards from Parkinson, the fossils expert of his youth, we find something very different. Buckland first seemed to have proved that Noah's Flood had happened straightforwardly as described in the Bible, drowning all hyenas but two. Even then, his recognition of much older strata required him to interpret the first verse of Genesis in a particular way, which affronted literalists. How far geological investigation and biblical interpretation should be combined or separated was a matter of controversy: but it was a controversy among clergymen, such as Buckland and Cockburn, rather than between believers and atheists. The discovery of much older forms of life, characterising epochs of immense length – and of the variable date of 'diluvial' deposits of gravel – meant that Buckland soon had to admit that the evidence from his cave did not relate to Noah's Flood.

But the consensus throughout the 1840s and 1850s was that the story in Chapter 1 of Genesis was verified by geologists in a general way and that while science and religion should not be mixed up, when properly understood they supported each other. Laymen like Davy, Murchison and Lyell were believers, though perhaps anticlerical or pantheistic: in this generation, there were very few infidels in the upper ranks of the scientific community. The Bridgewater Treatises, Buckland's in particular, had been successful.

Research and ritual

The first meeting of the British Association had been held in York in 1831,[62] under the auspices of the archbishop, Vernon Harcourt (whose son was the chief organiser) and despite the opposition of Dean Cockburn. Bones from Kirkdale and other local fossils were important and topical, but one of Buckland's children died, and to his chagrin he could not go. The next meeting was arranged for Oxford, and Buckland was the leading light: he managed to arrange for some important men of science, including John Dalton the Quaker, and Michael Faraday the Sandemanian, to be given honorary doctorates. But 1832 was the year of the Reform Bill, and conservatives deplored this gesture by the university towards Dissenters. In the subsequent Tractarian excitement, Buckland found himself (with other Oxford men of science) cold-shouldered, despite his friendship with Pusey, who became increasingly austere in the face of family tragedies. Everybody was too busy going to sermons, or discussing baptismal regeneration or the doctrine of reserve, to bother with Buckland's or other science professors' lectures. Younger enthusiasts seemed absorbed in ritual, later causing great public controversy.[63]

To most, science seemed opposed to religion only in that it took time, which should have been spent in church; but others looked for dogma, unquestioning faith in saintly miracles and clericalism,[64] and Buckland was disheartened. The venom and unfairness with which clergy attacked each other also appalled Charles Daubeny, chemist and Professor of Botany and Rural Economy at Oxford. He congratulated himself that as a layman it was not his vocation to take part in such conflicts[65] and resolutely defended the reputation of his mathematician colleague Baden Powell, who had died soon after contributing an essay on miracles to the liberal theological volume, *Essays and Reviews* (1860). For these, as for Buckland, Oxford from about 1833 to 1845 was thoroughly uncongenial.

Davy, pondering retirement while he worked on his dialogues, had considered that Robert Peel would be the ideal person to follow him as president of the Royal Society. Peel was an energetic politician, the Home Secretary who gave his name to the British bobby with his founding of the Metropolitan Police. He had been helpful to Davy over the founding of the London Zoo, and in providing government funds for Royal medals to reward scientific research. He went on instead to become a reforming prime minister and in 1845 threw Buckland a lifeline, nominating him to the position of Dean of Westminster (in succession to Wilberforce, made Bishop of Oxford partly to squash the Tractarians). A career in geology had led, in the way Huxley would later deplore, to an important position in the Church, in charge of the great national shrine and pantheon right beside the Houses of Parliament, where Davy and Lyell are commemorated as well as Darwin.[66] Clearly, Genesis and geology, God and nature, were not yet seen to be at strife.

CHAPTER 6

High-church science

DEFENDING THE EIGHTEENTH-CENTURY CHURCH of England against charges of complacency, torpor and indifference to education might seem a quixotic enterprise, not to be lightly or wantonly undertaken. On that point, if nothing else, the Evangelicals of the early nineteenth century and the Oxford high-church Tractarians a generation later seem to have agreed. Only when the broad-church *Essays and Reviews* was published in 1860 did the task of rehabilitating, or understanding, this part of the past begin.

In that scandalous collective volume[1] – which introduced biblical criticism to the great British public[2] just when the *Origin of Species* came out – Mark Pattison[3] wrote sympathetically on 'Tendencies of religious thought in England, 1688–1750' while Frederick Temple discussed 'The education of the world'. The ballyhoo from the clergy (with Soapy Sam Wilberforce promoting heresy charges against some authors) was so appalling that Temple remained somewhat anti-clerical even after he became Archbishop of Canterbury.[4]

If not exactly against general education, certainly eighteenth-century churchmen had been uneasy about educating foundlings, or indeed most children; and their successors were rather ashamed of it. It would have been a poorly educated nation at all levels, had it been left to the national church. England had only two universities – as had Aberdeen – until 1832, when Durham and then London received their charters. Dissenters could not attend Oxford or Cambridge without forswearing their beliefs, since attendance at college chapel and subscription to the church's Thirty-Nine Articles were both required. Well into the nineteenth century these universities were in effect seminaries, with over half their students aiming at ordination – for the others, they were finishing schools. There were scholarships, which enabled some very able boys to climb the ladder, as Cardinal Wolsey had, from cottage to high office – the church was in part a meritocracy – but these were exceptions. Educating every ploughboy would surely raise expectations that – it seemed to the conservative-minded – could never be fulfilled and must issue in frustration and social upset.

After all, until the eighteenth century there had seemed no good reason to think that there could ever be significantly more white-collar jobs: most people were going to have to work with their hands, on the land or in long-established crafts, or join the army or navy. Most boys were expected to follow their father's occupation: the exceptional ploughboy might get to university, and enter one of the learned professions – the church, the law or medicine – but the sons of clergy, lawyers and doctors would be denied their reasonable expectations if significant numbers from the lower orders were to be educated above their station.[5]

These doubts were not unreasonable. Right through the nineteenth century the professions were over-full in Britain: many doctors, like the young T.H. Huxley (he solved the problem by joining the navy), had trouble making ends meet and when Bishop Van Mildert,[6] founding the University of Durham, wrote to his fellow-bishops asking if they would ordain suitable Durham graduates, the Bishop of Rochester grumbled there were too many clerics already. As we know from literature, there were many impoverished curates with no hope of getting a comfortable 'living'. Until the Ecclesiastical Commissioners were set up after the Reform Bill of 1832, the church was a lottery with some plum prizes and many duds.[7]

Establishment and Dissent

We are familiar with the Unitarian Priestley's denunciation of established churches, and the separation of church and state urged by his Deistic friends Franklin and Jefferson; but, even so, it was possible to support establishment even if one were a cool and sensible sceptic[8] – a worldly wiseman rather than a keen churchman. Indeed, enthusiasm aroused alarm in the late eighteenth century, just as in our day many people dread fundamentalism and cults, and feel uneasy about 'alpha' courses, 'Toronto blessings', happy-clappies and charismatics.

With its formal liturgy, learned ministry and national position, an established church could canalise religious emotion into seemly and appropriate channels.[9] Seeing what was and is done in the name of religion, there was clearly much to be said for a respectable, if rather damp, Laodicean[10] and Erastian establishment in which piety and good works could be encouraged.

But an established church has to be broad, to incorporate as many believers as possible: Calvinist and Arminian, evangelical and Catholic.[11] It is a church based on parishes, rather than a 'gathered' church of like-minded people, so there has to be some fudging, with an emphasis on custom and procedure rather than chilly logic. As a result, most bishops in the Church of England were (and are) terrified of rocking the boat,[12] looking to achieve peace and compromise, even harmony if possible. In contrast, offsetting this timid conformity, vicars and rectors had assured income and tenure: they were almost impossible to remove, so they could and did behave and teach as they saw fit, even if their bishop or their flock didn't like it.

There was suspicion, and sometimes hatred, between church and chapel in

England right through the nineteenth century, and this is crucial in understanding Victorian politics. Church and chapel might unite in condemning Roman Catholics and collaborate in the British and Foreign Bible Society, but good relations were seen as worthy of comment. Rivalry was more typical, even in education and higher learning. It was Dissenters who took seriously the education of the young in the opening years of the nineteenth century, when the monitorial systems of Lancaster and Bell transformed elementary teaching. Their example was followed by the Church and its National Society, encouraged by evangelicals like Hannah More.

Education beyond schools

Dissenters were prominent in the expanding economy, and in education for its new openings: industry and commerce required managers, mechanics and clerks, and new professional groups like engineers and actuaries came into being. In great industrial cities, Newcastle-upon-Tyne for example, Unitarians were prominent in setting up Literary and Philosophical Societies to diffuse high culture among the new middle class[13] and in London the Unitarian manufacturer Samuel Parkes urged in his *Chemical Catechism* (1807) that such parents should teach their sons science to prepare them for careers in industry.

At Oxford and Cambridge in the 1780s, lectures on chemistry given by Dr Thomas Beddoes and by (the future bishop) Richard Watson had attracted huge audiences, though they were not part of the syllabus – which consisted of classics and an insular, Newton-worshipping kind of mathematics. But this chemistry was partly what continental Europe would have called cameralistics – preparation for grandees to administer estates with mineral riches[14] – and partly exciting stuff that any well-educated person ought to know about. Davy's amazingly successful lectures to the ladies and gentlemen at the Royal Institution in the 1800s were on the same pattern.

Paradoxically, the two authors and men of science we are now concerned with, William Swainson (a layman) and William Kirby (a clergyman) were high-churchmen, in a tradition going back to the non-jurors of 1689 who – believing in the duty of passive obedience, and prepared to resign from their benefices – refused to repudiate their oath of allegiance to James II. The tradition was continued in the eighteenth century by William Jones of Nayland,[15] an important figure in the Church, though (as a Tory) never given high preferment. The books of Swainson and Kirby, written in the 1830s, were addressed to the new reading public that had recently come into being.

The 1820s, with the revolution in book publishing brought about by steam-powered printing presses, cheap paper made from wood pulp or esparto grass, wood engraving and publishers' case-bindings, was the era of the March of Mind, or Intellect. The (Dissenters') Society for the Diffusion of Useful Knowledge competed with the (Church's) Society for the Promotion of Christian Knowledge in issuing little books for children and for adult readers, while the coming of gaslight made possible evening classes and late opening of libraries at Mechanics' Institutes. The new readers were serious: much of their reading matter was in small volumes, often duodecimo, with small type and narrow margins. These books contained a lot of information for the avid inquirer in what, to our generations, would seem a rather unpalatable form (though the embossed case-bindings are fun).

Henry Bohn was one publisher involved in this revolution; another was Thomas Longman, who published a large set of little volumes called *The Cabinet Cyclopedia*. The general editor, an efficient recruiter and a man of parts, was Dionysius Lardner. Some said sneeringly that his real name was Dennis; others called him 'The Tyrant' after Dionysius of Syracuse, because he pushed his authors so hard. He was famous for declaring, just before it happened, that steam navigation across the Atlantic would never be possible because the ship would have to carry so much fuel that there would be no room for cargo or passengers; and also because he eloped romantically to Paris with the wife of an army officer. The small octavo volumes in this series would together form an impressive non-fiction library; but they did not constitute an encyclopaedia and they were not textbooks.[16] They were aimed at those who wished to educate themselves, in this age of improvement and self-help. Among the authors Lardner recruited was William Swainson, who undertook to write numerous volumes on natural history.

William Swainson

Swainson[17] was born in Liverpool in 1789, the son of a customs officer. His father naturally expected him to work in the customs house also, but William developed a taste for travel and natural history that led him instead to join the Commissary of the Army. In 1807 he was sent to Malta; after the war ended in 1815 he elected to go on half pay. His time in the army, and his conviction that he belonged to a distinguished family that had gone down in the world, made him prickly and hard to deal with. He found it hard to keep a friend and his life was punctuated with furious rows: 'bred up with somewhat of aristocratic notions, and accustomed, when on service, to command rather than obey, I had a rooted dislike of all commercial affairs, and would rather have gone once more on active duty than have sat behind a desk'.[18] He also suffered, like Priestley, from a speech impediment; as a result, his direct and offensive remarks were put down on paper, where they would be harder to forgive or forget.

In 1816 he set out for Brazil, but unfortunately there was a war on: he barely got ashore and his collections of specimens were sparse.[19] Nevertheless, on returning to Britain he was elected FRS; and he got to know William Leach,[20] in charge of zoology at the British Museum, who encouraged him to take up zoological illustration using the new art of lithography. In this, a drawing was made on a suitable stone with a wax crayon: the stone was wetted, and then inked with an oil-based ink, which adhered to the wax but not to the wet stone, so that a print could be made. This was much cheaper than engraving on copper, and moreover the artist could be in charge of the whole process, rather than having his drawing translated by a craftsman into the less exuberant language of engraving.

Swainson, a very talented artist, became the first person in Britain to publish works of natural history (on birds and shells) illustrated by lithographs,[21] and he employed the system of publishing in parts. What subscribers paid for part 1 would pay for the printing of part 2, and so on, which benefited Swainson; and the overall price for what were superbly illustrated books seemed less daunting to readers, if spread over a year or two as the parts came out. The plates were hand coloured. The 'pattern plates' coloured by Swainson for his team of colourists are at the Linnean Society in London, where there is also a splendid and very revealing archive of his correspondence.

When in 1822 Leach retired from the British Museum, Swainson hoped to become his successor: but the post went to J.G. Children, down on his luck after the failure of his bank. He was a chemist rather than a zoologist, but he was also a friend of Davy's – and Davy, as President of the Royal Society, was a trustee of the museum. Disappointed, and thenceforward a grumbling and disgruntled man, Swainson found that his inheritance from his father was scanty, and he would have to support himself and his family by pen and pencil.

Swainson goes his own way

In 1833 he signed a contract with Longman to write fourteen volumes on natural history, of three hundred pages each, to be illustrated with wood engravings. (Lithography worked only for short runs, at the upper end of the market.) They were to be produced for £200 each at the rate of one every three months, for Lardner's series, and they illustrate why he was called a tyrant. Swainson had been working for Longman on an aborted encyclopaedia, and thus had some material to hand, but even so the contract was wholly unrealistic; volumes nevertheless steadily if slowly appeared from 1834.

The first was a *Preliminary Discourse*, for which the model was John Herschel's on natural philosophy in the same series – an accessible classic in philosophy of science.[22] At the outset, two features of it deserve notice as a statement about scientific and religious allegiance. Firstly, on the engraved title-page, where Herschel had a bust of Francis Bacon as founder of inductive philosophy, Swainson put Aristotle, whom he considered the founder of zoology and the true discoverer of Cuvier's law of the correlation of the parts of animals (which led to the reconstruction of fossils).[23] Lardner had written advising no bust, and especially not one of Aristotle or John Ray, but Swainson did not mind standing up to tyrants and one cannot but admire him for it.

Secondly, opposite the title page is a quotation (Herschel had again used

Bacon, on man as the minister and interpreter of nature) from Jones of Nayland:

> The world cannot show us a more exalted character than that of a truly religious philosopher, who delights to turn all things to the glory of God; who in the objects of his sight, derives improvement to his mind; and in the glass of things temporal, sees the image of things eternal.[24]

Using the last sentence, we can explore how those in the high-church tradition might differ from Dissenters in their attitude to the natural world. Bacon had popularised the idea that there were two books, the Bible and Nature ('There is a book, who runs may read', as Keble expressed it), which together told us about the infinite wisdom and goodness of God. All Christians from whatever denomination might unite in that task of natural theology, and many did so: but as we know, there were problems. Natural religion might come to seem sufficient to stand on its own. In that case, revealed religion, which ought to be the core of Christianity, would be merely a bolt-on extra available for those who required it – as indeed it is in the Deist William Wollaston's *Religion of Nature Delineated* (1722 and later editions).

Paley in his *Natural Theology* (1802) had revitalised the old idea of the world as a clock, and God the clockmaker; but the utilitarian philosophy essential to his rhetoric did not appeal to high-churchmen. They would have agreed with some other orthodox Christian thinkers that arguing for God's existence and goodness (applauding Him for a good bit of design, and giving Him the benefit of the doubt over general happiness) was impertinent and presumptuous, and the wrong way to go about instilling faith – meaning real belief and trust in a personal God rather than notional assent to various propositions.[25] But they also emphasised sacraments and hierarchy, seeing the world as a glass on which one should not stay one's eye, but look through and spy the heavens. For Swainson, the natural world was full of types and symbols: he noted that the chrysalises of stinging caterpillars point downwards towards Hell, for example – a point which the evolutionist A.R. Wallace in his copy[26] characterised first as 'mildly fanciful' and later as 'going too far beyond common sense'.

Going beyond common sense was something Swainson was seen as doing in the system of taxonomy that provided the organising principle for his books in Lardner's *Cabinet Cyclopedia* series. This was based upon trinities of circles where extremes met: at every level (species, genera, orders) there are three circles, which include everything. These are labelled the Typical, the Subtypical and the Aberrant: this last one is itself divided into three little circles. The two big circles and three little ones make five, and so the system was called Quinary – which disguises its basis in threesomes. Swainson did not invent this system, but became its apostle. The Great Chain of Being, a ladder from amoebas up through all the links to mankind (and perhaps on to angels and archangels), had been abandoned in the light of Cuvier's work; and animals were therefore classified in natural groups on Aristotelian lines, in an overall untidy pattern looking like a bushy, shrubby tree with four main stems.

The quinary system with its levels provided instead an elegant and intelligible framework, and might even predict undiscovered creatures: it seemed a real key to the natural world (as the Periodic Table was later to be in chemistry). It aroused the interest not only of Wallace, but also of T.H. Huxley and Charles Darwin, who were fascinated but then repelled by so static and dogmatic a scheme; which also relied on subjective judgements of resemblances and affinities, rather than careful weighing of evidence from dissection as well as appearance and habits.

The system, denounced as visionary and procrustean at the British Association in 1840 and 1843 by Hugh Strickland,[27] lingered on in the Antipodes, whence Swainson had emigrated, and in Canada;[28] but in Britain it expired, and it never caught on in France or Germany. Swainson had undertaken to write a lot of books, on a wide range of topics, fast; and his Trinitarian circles gave them a much-needed if artificial theme.

Readers of Swainson's little volumes would thus learn a lot of natural history, within a framework which (like Linnaeus') was based largely upon external characteristics – thus Swainson put the marsupial 'Tasmanian wolf' with the dogs rather than (like Cuvier, who took anatomy and physiology into account) in a quite different group near the kangaroos. The different books cover the geographical distribution of animals, and the lives of eminent naturalists (Swainson giving himself the longest entry); and include conchology[29] (on which Swainson was an expert, and had done sumptuous lithographs) as well as vertebrate zoology.

His *Preliminary Discourse* (1834) was more personal (not to say idiosyncratic) than Herschel's, and thus while it may well have kindled enthusiasm it never became a classic – it contains praise for Adam Sedgwick and his defence of natural theology, and urges more support for natural history from government and universities: in the controversies of the day, he was a 'declinist' believing with Babbage that British science was going downhill. Swainson's own grumbles come through in various places, notably in a footnote where he compares his own treatment with that of Bavarians who had explored Brazil (but rather more thoroughly, though he does not say so).[30]

William Kirby

In its search for hierarchical order, its delight in symbol and type, and its Trinitarian taxonomy, Swainson's series has a distinctively Church and Tory character. The same is true of William Kirby's Bridgewater Treatise[31] on the history, habits and instincts of animals as showing the wisdom and goodness of their Creator. Most of his examples, as we would expect from an entomologist, are invertebrates – illustrated by lithography in the first edition (1835) and woodcuts in 1853 – although men, birds and whales do feature, and the second volume of the 1853 edition has an attractive frontispiece of platypuses at play. The Treatises were works of popularisation, science made palatable (and indeed momentous) by natural theology; and, like Paley, the authors were aiming to bring their readers conviction, rather than logical proof. Aimed at a different and, it seemed, less numerous class of readers than Lardner's, the volumes were large and handsomely-printed octavos, modestly bound in cloth cases with paper labels. The aim was to broaden the education of the educated; but their readership must have overlapped with Lardner's, and the 1853 edition, published by Bohn, looks very like the *Cabinet Cyclopedia*.

Kirby shows great respect for both Aristotle and Bacon, and duly mounts an attack on Lamarck (where he explicitly follows Lyell) and Laplace for their evolutionary speculations. He too refers to Jones, his 'venerated friend', and remarks[32] of natural philosophy and the scriptures:

> The Bible was not intended to make us philosophers, but to make us wise unto salvation. But it does not follow, because we seek for religious truth principally in the Bible, that we can derive none from the study of natural objects; nor, on the contrary, because we are not to go to the Bible for a system of philosophy, that no philosophical truths are contained in it.

Elsewhere he remarks 'that in order rightly to understand the voice of God in nature, we ought to enter her temple with the Bible in our hands'. Passages in the Bible concerned for example with Cherubim are referring to the forces and powers of the Newtonian world.[33]

Animal metamorphoses are, as in ancient tradition, compared to death and resurrection; and Kirby, in fine polysyllabic style, discusses hair, which in scripture (we are told) stands for power and probably conducts subtle fluids (such as heat or electricity) in and out of animals:

> and thus the various piligerous, plumigerous, pennigerous, and squamigerous animals, may offer points and paths, not only to the air, but to more subtle fluids, either coming or going, whose influences introduced into the system, may add a momentum to all the animal forces, or which, having executed their commission become neutralized, may thus pass off into the atmosphere.

Since God did nothing in vain, we might be assured that the 'system of *representation* was established with a particular view'. All things in the world are significant, and our great Instructor's creatures are placed before us as signs or symbols representing other things, as well as playing their part in the general drama. Since the Fall, learning by Revelation rather than through symbols and emblems has become more important; but spiritual truths are 'reflected as by a mirror, and shown as it were enigmatically', making the study of nature of the first importance.

Especially we find examples of benevolent and beneficent, or malevolent and mischievous, animals that represent the two classes of spirits to be found in the invisible world. Animals are not only directly useful for us, but also emblematic: we are here in the realm of Aesop's fables and the bestiaries, and yet informed by up-to-date science. Many readers of Kirby must have been puzzled by the mixture of old and new, but the series was directed particularly at the clergy and there was much in the book that could be used in sermons: this was not just information about surprising facts of nature.

Kirby especially refused to do what was expected of Bridgewater authors, to separate the evidence of nature for God's existence, wisdom and benevolence from the evidence of revelation. Nobody reading his book could have supposed that zoology could lead to some natural religion: it only fitted into a big picture when associated with the Bible. The Book of Nature and the Bible illuminated each other and should not be separated. High-church science may be a surprising category, but we do find it in these educational publications; and it is rather different from Rational Dissent as in Priestley and his disciples, from Paley's rhetoric of evidences, and also from Newman's sceptical unease about all natural theology.[34]

High-church science like Swainson's and Kirby's had by the 1840s come to look old fashioned. Beyond the sphere of orthodox Oxford and Cambridge men, those seeking the sublime in science turned towards a pantheistic spirituality, worshipping nature.[35] One reason perhaps why the Scientific Revolution (if there was one)[36] had happened in the West was that Nature was seen as the Creation,[37] as a comprehensible 'It' rather than an unknowable 'She'. Thus Robert Boyle had as a good Protestant denounced what he saw as Platonising attempts to revive the notion of Nature as a demiurge, doing God's bidding as best she could.[38] Admiring the great clock at Strasbourg, he popularised the notion of God the clockmaker; this idea[39] was revived with John Harrison's marine chronometers,[40] eventually so compact that they looked like big pocket watches. Paley's mechanical vision of the world as the best of all possible watches (with people and animals like watches too), in which all the parts worked smoothly together to produce the greatest happiness of the greatest number,[41] updated Boyle's conception in elegant prose.

Awesome science and Humphry Davy

Paley was an older contemporary of the Romantic generation – William Wordsworth, S.T. Coleridge and Walter Scott – and with the Romantic Movement, worship of Nature returned,[42] perhaps in the form of the Higher Pantheism, as in Alfred Tennyson's later poem of that name:[43]

> The sun, the moon, the stars, the seas, the hills and the plains –
> Are not these, O soul, the Vision of Him who reigns?
> Is not the Vision He? tho' He be not that which He seems?
> Dreams are true while they last, and do we not live in dreams?
> Earth, these solid stars, this weight of body and limb,
> Are they not sign and symbol of thy division from Him?
> Dark is the world to thee: thyself art the reason why;
> For is He not all but that which has power to feel 'I am I'?
> Glory about thee, without thee; and thou fulfillest thy doom
> Making Him broken gleams, and a stifled splendour and gloom.
> Speak to Him thou for He hears, and Spirit with Spirit can meet –
> Closer is He than breathing, and nearer than hands and feet.
> God is law, say the wise: O Soul, and let us rejoice,
> For if He thunder by law the thunder is yet His voice.
> Law is God, say some: no God at all says the fool;
> For all we have power to see is a straight staff bent in a pool;
> And the ear of man cannot hear, and the eye of man cannot see;
> But if we could see and hear, this Vision – were it not He?

Tennyson's poetry was much admired by Thomas Henry Huxley,[44] who concluded his Royal Institution lecture on *The Origin of Species* in February 1860 with a quotation from the *Idylls of the King*, and believed that *In Memoriam* was a good example of scientific method. These two Victorian apostles, of doubt and of agnosticism, were thus not very far apart; and Huxley made sure that the Royal Society was officially represented at Tennyson's funeral.

It would be a great mistake to think that nineteenth-century scientists were unaffected by, or unsympathetic to, Romanticism:[45] rather they were heirs of both the Enlightenment and the Romantic Movement, and this is visible in their spirituality. Thus the young Humphry Davy,[46] experimenting in the 1790s in Bristol with laughing gas and other drugs, had under Coleridge's tuition a 'distinct sympathy with nature', when 'Every thing seemed alive, and myself part of the series of visible impressions' so that he would have felt pain in tearing a leaf from a tree.[47] Asked by Wordsworth as a sympathetic spirit to read the proofs of the second edition of *Lyrical Ballads*, he punctuated them for the press; and among the poems he wrote himself is a rhapsody of uncertain date:

> Oh, most magnificent and noble nature!
> Have I not worshipped thee with such a love
> As never mortal man before displayed?
> Adored thee in thy majesty of visible creation,
> And searched into thy hidden and mysterious ways
> As Poet, as Philosopher, as Sage?

HIGH-CHURCH SCIENCE

Invited to London as lecturer at the newly founded Royal Institution, Davy began his course in January 1802 with an inaugural lecture that made him a feature of the London scene. Men and women of fashion flocked to hear him,[48] and on lecture nights a one-way traffic system was inaugurated to reduce congestion in Albemarle Street where he held forth. Calling attention to scientific progress and its unlimited future possibilities and practical applications – he can be seen as the high priest of applied science[49] – he went in for sexy rhetoric:[50] 'Not content with what is found upon the surface of the earth, [the chemist] has penetrated into her bosom, and has even searched the bottom of the ocean for the purpose of allaying the restlessness of his desires, or of extending and increasing his power'. As we shall find, searching into Nature's hidden and mysterious ways might involve worship, but it was also macho: indeed the rhetoric of science has often involved images of warfare and aggression, of possession, as though all knowledge were carnal.

There is also some Baconian rhetoric of torture: the man of science is a 'master, active with his own instruments'. There is nothing like a bit of such aggression or lust to bring a lecture to life; and this was safely pre-Victorian. Davy also had a more feminine imagery, which we are apt to think more appropriate to religion, of discovery regularly visiting the man of science. He did believe that humility of a kind was essential. Science required an unwillingness to jump to conclusions and a readiness to drop preconceptions, as in the Baconian tradition, where science was refined common sense rather than the arrogant mathematical pursuit of recondite necessary truths.

Davy and his generation were living through a prolonged world war (1794–1815, with a brief truce in 1802) that had begun when the French began to export their Revolution of 1789, backed first with terror and then with the Grande Armée. Even if scientific ideas lay behind the revolution in France, it had taken such a different turn from the American model that natural philosophers in Britain felt it showed that science rightly understood did not entail materialism and atheism. The French called for liberty, equality and fraternity: liberty was no problem, for it was the birthright of every Englishman; equality was for Davy and his audience absurd, because the world was clearly unequal, and science was intertwined with capitalism; but the scientific community could perhaps be a fraternity. Not a sorority, though: for women, science would be a spectator-sport. It was essential for Davy and his contemporaries to show that in a free country, where religion and law were respected, science could flourish just as well as in its world centre, Paris: he was fighting the same war as Nelson and Wellington, but on the intellectual front, in proving that chemistry depended on electrical force rather than matter.

At the end of his life, prematurely old at forty-nine after a stroke, visiting Italy and the Alps in search of health, Davy wrote his little book of dialogues – essentially about science and spirituality[51] – *Consolations in Travel*. It begins with a reverie in the Colosseum, in the manner of Edward Gibbon; and transport to the sphere of Saturn, the seventh heaven, in the manner of Saint

Paul. There the author obtains a progressive vision of human history; and an assurance of immortality and reincarnation in progressively more æthereal bodies as we wear out our present machinery. Because he was writing dialogue, Davy could propose various doctrines and ideas; but the general message is clear. Like his friends Coleridge and Wordsworth in their great Odes (on Dejection, and Immortality), he was coming to terms with age, the associated loss of creative imagination and death.

Davy's family was Anglican; and after a period of youthful materialism he returned to religion, but in what seems an impersonal and pantheistic form,[52] writing in *Consolations*:

> The doctrine of the materialists was always, even in my youth, a cold, heavy, dull and insupportable doctrine to me, and necessarily tending to atheism. When I had heard with disgust, in the dissecting rooms, the plan of the physiologist, of the gradual accretion of matter and its becoming endowed with irritability, ripening into sensibility and acquiring such organs as were necessary, by its own inherent forces, and at last arising into intellectual existence, a walk in the green fields or woods by the banks of rivers brought back my feelings from nature to God; I saw in all the powers of matter the instruments of the deity; the sunbeams, the breath of the zephyr awakened animation in forms prepared by divine intelligence to receive it; the insensate seed, the slumbering egg, which were to be vivified, appeared like the new born animal, works of a divine mind; I saw love as the creative principle in the material world, and this love only as a divine attribute. Then, my own mind, I felt connected with new sensations and indefinite hopes, a thirst for immortality.

'Atheism' still implied, as it had in the seventeenth century, ruthless hedonism – a way of life rather than a speculative system.

But there was indeed materialistic and evolutionary medical teaching in London by 1830,[53] long before *Vestiges*[54] (1844), Darwin's *Origin of Species* (1859), or Tennyson's famous stanza from *In Memoriam* (1851) on man, Nature's last work:[55]

> Who trusted God was love indeed
> And love Creation's final law –
> Tho' Nature red in tooth and claw
> With ravine, shriek'd against his creed.

What his religion brought Davy was thus belief in design and in benevolence, coupled with a feeling for divine inscrutability that made him impatient with Paleyan reasoning; and it brought him a belief in immortality, especially important to a Romantic genius.

On isolating potassium in 1807, Davy had danced about the laboratory in ecstatic delight; joy in nature and in science, as well as awe, were appropriate to Nature's votary: in the laboratory of his great rival, J.L. Gay-Lussac dancing after success in the laboratory seems to have been a little calmer and more liturgical.[56] There were exciting dangers too in the manly activities of the laboratory: Davy was nearly killed by respiring carbon monoxide, and was injured in explosions. And her worshipper could become a benefactor to humanity, as Davy proved with his work in agriculture and the safety lamp for miners (1815), which made him one of the most famous men of his day. He had begun to fulfil Bacon's promise, that through science the consequences of the Fall would be mitigated if not removed

Higher pantheism

Unitarianism had been a feather-bed to catch a falling eighteenth-century Christian;[57] for Davy – and, as the nineteenth century wore on, for others – pantheism filled that role. It had the advantage of not requiring attendance at public worship on Sundays, or having to listen to sermons: nature could be adored in the green fields, with rod and line beside a trout or salmon river, in the sublimity of Alpine summits and passes, at great scientific meetings or in the laboratory.

The doctrine was loose and accommodating; it pointed towards the Church Scientific, and the New Reformation, of T.H. Huxley,[58] where men of science would be the new clerisy, presiding over an educated and responsible laity, and dealing with plagues (God's punishments for ignorance and laziness) through waterworks and sewage systems rather than prayer. In the shorter term, it enabled devotees to distance themselves from organised religion, with its dogmas, anathemas, denominations and educational systems based on catechisms and classics. Davy, the young Huxley and other self-made scientists working in London saw their colleagues in Oxford and Cambridge (with their religious tests) as muzzled by a party line; and were correspondingly attracted to a loose and personal religion without a personal God.

Davy's greatest pupil was Michael Faraday; who came similarly from a humble background, but whose character, outlook and ambitions were very different from Davy's. He had no time for mainstream churches, being a Sandemanian:[59] a small sect, now extinct, of biblical literalists, without clergy, they did their best to keep themselves unspotted by the world. Faraday thus avoided high office in scientific institutions, and did not dine out unless he felt it was his duty to do so. As an elder in the church, he preached on Sundays; and was impeccably orthodox in his religious beliefs. He was quite unlike Davy, who enjoyed hobnobbing with the mighty and was happy to accept responsibility within the scientific community; and who grumbled at the end of his life that he had not had sufficient honour and recognition.

Faraday would have had little sympathy with the natural religion that Davy admired, in his lines written at Tivoli:[60]

> Thy faith, O Roman! was a natural faith,
> Well suited to an age in which the light
> Ineffable gleam'd thro' obscuring clouds
> Of objects sensible, – not yet revealed
> In noontide brightness on the Syrian mount.
> For thee, the Eternal majesty of heaven
> In all things lived and moved, – and to its power
> And attributes poetic fancy gave
> The forms of human beauty, strength, and grace.
>
> . . .
>
> I wonder not, that, moved by such a faith,
> Thou raisedst the Sybil's temple in this vale,
> For such a scene was suited well to raise
> The mind to high devotion, – to create
> Those thoughts indefinite which seem above
> Our sense and reason, and the hallowed dream
> Prophetic. – In the sympathy sublime,
> With natural forms and sounds, the mind forgets
> Its present being, – images arise
> Which seem not earthly, – 'midst the awful rocks
> And caverns bursting with the living stream, –
> In force descending from the precipice, –
> Sparkling in sunshine, nurturing with dews
> A thousand odorous plants and fragrant flowers.
> In the sweet music of the vernal woods,
> From winged minstrels, and the louder sounds
> Of mountain storms, and thundering cataracts,
> The voice of inspiration well might come.

Davy's personal beliefs were painfully worked out in his last wanderings, when he found himself spared long enough to look for a meaning for his life, and to move from science towards wisdom.

As well as Faraday, he had a fictional pupil: Victor Frankenstein. Mary Shelley's model for Professor Waldman,[61] Frankenstein's teacher, was Davy, whom she had known in London. Like Davy, he declared of chemists that 'they penetrate into the recesses of nature, and shew how she works in her hiding places. . . . The labours of men of genius, however erroneously directed, scarcely ever fail in ultimately turning to the solid advantage of mankind'. A sorcerer's apprentice who fails to mother his ill-begotten offspring, the optimistic Frankenstein is transformed from victor to victim; and the book is about the folly of playing God, the need for moral responsibility in science, and the ambiguities of progress.

Tennyson also doubted science and progress as much as he did religion: for example in his two great poems on Locksley Hall. In the first, 'Forward, forward let us range,/ Let the great world spin for ever down the ringing grooves of change' is set against life in some timeless tropical Eden; and, in the later poem, 'Evolution ever climbing after some ideal good,/ And Reversion ever dragging Evolution in the mud' balance each other out. We end with what should be a message of hope, 'Love will conquer at the last'.[62] But there is no place here for confident faith, and finding a meaning to life is a painful and uncertain process, involving golden and leaden echoes, never finished until death.

Some men of science found the sublime when they took the wings of the morning,[63] and actually ascended into the heavens in a balloon. Thus the eminent astronomer and meteorologist James Glaisher wrote:[64]

> I have experienced the sense of awe and sublimity myself, and have heard it on all sides from aeronauts, who have both written and said the same. For my part, I am an overwrought, hard-working man, used to making observations and eliminating results, in no way given to be poetical, and devoted to the immediate interest of my pursuit, and yet this feeling has overcome me in all its power. I believe it to be the intellectual yearning after the knowledge of the Creator, and an involuntary faith acknowledging the immortality of the soul.

Glaisher's interests were very wide; but by the 1860s science was becoming a specialised activity, with undergraduate degrees available in its various branches. The position of scientists in Britain was much stronger than it had been in Davy's day, and in 1870 it received a boost from the Franco-Prussian War, where the more scientific nation won – the effect was comparable to that of Sputnik in 1957. But science might still be seen as materialistic, with worries now about German influences; or as a merely technical activity, 'normal science', in which important questions were simply not raised. This would be utterly unspiritual as well as dispiriting, and would not measure up to the standard of a liberal education suitable for a gentleman.

Questions that no experiment can answer

John Tyndall, successor to Davy and Faraday at the Royal Institution and much influenced by Thomas Carlyle and his epic view of history,[65] took it upon himself to deal with both these objections. Through the nineteenth century, Davy's view that science was a creative process gained increasing favour among intellectuals,[66] and later came to be associated with Whewell's hypothetico-deductive idea of method. In 1870 Tyndall addressed the British

Association for the Advancement of Science on 'The Scientific Use of the Imagination'. He ended with an encomium upon geologists in particular, but scientists in general:[67]

> Their business is not with the possible, but with the actual – not with a world which might be, but with a world that is. This they explore with a courage not unmixed with reverence, and according to methods which, like the quality of a tree, are tested by their fruits. They have but one desire – to know the truth. They have but one fear – to believe a lie. And if they know the strength of science, and rely upon it with unswerving trust, they also know the limits beyond which science ceases to be strong. They best know that questions offer themselves to thought, which science, as now prosecuted, has not even the tendency to solve. They have as little fellowship with the atheist who says there is no God, as with the theist who professes to know the mind of God. 'Two things,' said Immanuel Kant, 'fill me with awe: the starry heavens, and the sense of moral responsibility in man.' And in his hours of health and strength and sanity, when the stroke of action has ceased, and the pause of reflection has set in, the scientific investigator finds himself overshadowed by the same awe. Breaking contact with the hampering details of earth, it associates him with a Power which gives fulness and tone to his existence, but which he can neither analyse nor comprehend.

Reverence and awe were words used by Davy nearly seventy years earlier: Tyndall's faith in science was very strong, but so was this pantheism.[68]

Because he was an ally of Huxley, and because in 1874 he declared at another British Association meeting, in Belfast when he was president, that men of science claimed, and would wrest from theologians, the entire domain of cosmological theory, he is often counted among agnostics or scientific naturalists;[69] but this is not quite true. Historians might think that as Victorian Britain could be described by Germans as the land without music, from its dearth of eminent composers; so it could be called the land without theology. Who then were the theologians in Tyndall's sights? It seems again as if he means denominational spokesmen, quarrelling and pontificating at a time when religion and politics were intimately intertwined in Britain; and that pantheism could distance its believers from such unseemliness. *Odium theologicum* is a longstanding disease, and outsiders might well marvel at that time how these Christians hated each other.[70]

Tyndall was an intrepid mountaineer, with several first ascents to his credit: his ice axe is preserved in the museum at Zermatt in Switzerland, and a subpeak of the Matterhorn is named after him. In rhetoric owing something to Alexander von Humboldt[71] he described for example the night before the first ascent of the Weisshorn:[72]

> An intensely illuminated geranium flower seems to swim in its own colour which apparently surrounds the petals like a layer, and defeats by its lustre any attempt of the eye to seize upon the sharp outline of the leaves. A similar effect was here observed upon the mountains; the glory did not seem to come from them alone, but seemed also effluent from the air around them. This gave them a certain buoyancy which suggested entire detachment from the earth. They swam in splendour, which intoxicated the soul, and I will not now repeat in my moments of soberness the extravagant analogies which then ran through my brain.

Upon the summit, he opened a notebook to make a few observations, but 'soon relinquished the attempt. There was something incongruous, if not profane, in allowing the scientific faculty to interfere where silent worship was the "reasonable service".' Tyndall sometimes took friends mountaineering, but usually climbed with one or two local guides – forgotten heroes, like the assistants who did so much research for great Victorian scientists.[73] Climbing was not wholly unlike communing with nature in the laboratory. And sometimes he preferred to be alone,[74]

> though the right to do so ought to be earned by long discipline. As a habit, I do deprecate it; but sparingly indulged in, it is a great luxury. There are no doubt moods when the mother is glad to get rid of her offspring, the wife of her husband, the lover of his mistress, and when it is not well to keep them together. And so, at long intervals, it is good for the soul to feel the influence of that 'society where none intrudes'. . . . The peaks wear a more solemn aspect, the sun shines with a more effectual fire, the blue of heaven is more deep and awful, the air seems instinct with religion, and the hard heart of man is made as tender as a child's. In places where the danger is not too great, but where a certain amount of skill and energy are required, the feeling of self-reliance is inexpressibly sweet, and you contract a closer friendship with the universe in virtue of your more intimate contact with its parts.

Tyndall's writings about the mountains are masterpieces in their way, full of feeling and eloquence that must have made him a worthy successor to Davy and Faraday as a public lecturer on science. But though pantheistic experiences can be thus described, they cannot be readily shared. A congregation making its way up towards the Matterhorn could not have felt like Tyndall, alone with his own solitariness, making friends with the universe amid the sublime majesty of creation.

Davy, Faraday and Tyndall all escaped to the Alps, or to other mountains and wildernesses, for refreshment (and to get fit) after hard work in London. By

Tyndall's time, the Alps in the summer holidays were becoming full of writers and intellectuals, professors and professional men; and often their wives too, who used mules where we would use cable-cars. The spiritual fuel there acquired might keep a pantheist going for several months, as it did with Tyndall; but pantheism is probably not a faith that makes it easier to confront the quotidian and the humdrum. Davy used it in facing lonely and premature old age, but the less privileged could not aspire to long foreign holidays. Nevertheless, pantheism seems to have played an important role as an expression of élite scientific spirituality for those who found the churches uncongenial. No doubt it still does, though perhaps modern physics has made Nature even more of an unknown god than she seemed to Davy, Glaisher or Tyndall.

CHAPTER 7

God working His purpose out?

THE NINETEENTH CENTURY WAS THE great age of history as well as science. Church time is circular: seedtime and harvest, Christmas, Lent, Easter and Whitsun keep coming round. But time clearly has a direction: we grow up and we get older. The Bible shows God's dealing with His people over centuries. Is such linear time just one thing after another? Historians, notably those associated with Romanticism, came to appreciate that the past was a foreign country, where they did things differently – though only in 1953 did the English novelist L.P. Hartley supply those words to clothe the thought. Imagination was required, as well as the careful study of documents and material objects, to understand the past. Historical imagination began to be valued, just as scientific imagination was: antiquarians and chroniclers turned themselves into historians with more or less gripping narratives to tell. Walter Scott and other romantic writers aroused real enthusiasm for history and its sublime perspectives.

To many in the eighteenth and nineteenth centuries, history (in its big picture) appeared straightforward and progressive: 'You can't stop progress' became proverbial. It seemed there really was a time-line rather like Priestley's, but some nations (and even parts of them) had got further on than others. The laggards, 'backward' peoples, might well find themselves inexorably displaced by the more advanced. Tasmanians, Africans, native Americans and others (even European colonists in Australia and Brazil)[1] found themselves at the receiving end of science and technology, and would no doubt have seen things differently from those writing the winners' history that we can readily read. After the experiences of the twentieth century, we may find it harder to see progress so simply: while few would want to forego all modern conveniences, we recognise what a contingent and branching chain history has followed, taking individuals and peoples down pathways that would not have been predicted.

Which way are we going?

Science may be deterministic, as Maurice Hare explained in 1905:

GOD WORKING HIS PURPOSE OUT?

> There once was an old man who said, 'Damn!
> It is borne in upon me I am
> An engine that moves
> In determinate grooves,
> I'm not even a bus, I'm a tram.'

Our future, though, is open: we are not on a tramline. On the way, Hitler may hijack the bus, people may get run over – there are good and bad effects for all of us. History did not always seem like that: in the nineteenth century especially, manifest destiny, civilising mission, or bearing the white man's burden seemed obvious. History had an onward and upward trend, and its high point (so far at least) was in Europe and North America and their inhabitants – though some looked back with longing to the medieval world.[2]

In the Bible too, history had its direction. God's action, working His purpose out, was visible right through the story of the Children of Israel. They made a covenant with God: often they broke it and were punished, often they repented and were restored, and finally they were dispersed by the Romans. God's spirit was seen at work in human history, rather than in the celestial mechanisms, well-adapted animals or sublime views that appealed to nineteenth-century men of science.

Belief in God meant something very different in these different contexts: to the ancient Israelite, who did not doubt that God existed, it meant trusting His promises declared by the prophets, and keeping His law and covenant (resisting the claims of all other gods). By the nineteenth century, especially in circles that were at all scientific, it meant adopting God's existence as an explanatory hypothesis for the order and beauty found in the world. To the future Cardinal Newman, the belief of the Israelites counted as real assent, God as an explanation was merely notional assent;[3] clearly they were not the same kind of belief. The second might easily be a mere 'God of the gaps'; certainly it seems – as Kirby, Whewell and others feared – far from a genuinely religious kind of thinking and feeling.

Charles Lyell had proposed a very long, but un-directional or un-teleological, history (or prehistory) of the Earth. Like earlier histories, his assumed that the forces acting had been always the same, and that the world had never been significantly different in the remote past: 'What has been will be again, what has been done will be done again; there is nothing new under the sun'.[4] Paley's Newtonian world had been like that.

Buckland's story was different. He did see a direction: a cooling Earth was being prepared for us to live on, and at different stages of its history God had created plants and animals appropriate to the conditions and to the end He had in view – such as having coal and iron deposits ready when we needed them. Buckland's was a world which, under God, was developing, indeed progressing because mankind was clearly of more interest to God than dinosaurs, though He had loved them too.

God had not completed His work sometime in the past, and He was not still enjoying His Sabbath-day repose as Deists had supposed: we had to see Him acting within ongoing processes, in change as much as in stability. Man had been placed upon the Earth when it was ready: creation was not something that happened over a short time in the past, but occupied aeons. God's purposes might gradually become evident to those with eyes to see them: His timescale was different from ours.

Behind this lurks the question whether the 'Watchmaker' is farseeing, as Buckland urged, or is blind. It seems safe to say that most evolutionists who preceded Charles Darwin and his *Origin of Species* (1859) did not see nature as blind, pouring forth from her lap, without discernment or parental care, her maimed and abortive children – as Hume had put it a century earlier.[5] They perceived progressive change proceeding from a largely benevolent Nature, if not from God.

Well versed in botany

Charles Darwin's grandfather Erasmus Darwin was a member – with Priestley, Watt and Wedgwood – of the Lunar Society of Birmingham[6] and about 1790 was regarded as England's major poet as well as an excellent doctor. His public reputation came from a series of didactic poems. We do not expect to learn facts from verses, but Lucretius had put his atomism into poetry, Alexander Pope's *Essay on Man* was a classic exposition of Enlightenment, and Victorians later learnt Latin grammar via a series of jingles as mnemonics.

Darwin's verses were unusual in requiring footnotes and endnotes, not normally expected of poetry; but readers could enjoy the language and the conceits, and not bother with the footnotes or scientific backing, if they wished – in fact, he wrote well in verse and prose. Erasmus Darwin made his reputation as a poet with *The Botanic Garden*, published in two handsome quarto volumes illustrated with engravings (including one by William Blake, after Henry Fuseli).

Curiously, the second volume, *The Loves of the Plants*, appeared (in 1789) before the first, *The Economy of Vegetation* (1791).[7] The second volume had depended upon playing with Linnaeus' sexual system: the very names of the classes indicated polygamy or polygyny, and Darwin played with the old idea of the language,[8] and the new idea of the loves, of the plants. Sexual innuendo was charming and made botany attractive rather than dryly systematic; and those wanting scientific botany could find it in the notes. Such lines as these about the mimosa:[9]

> Weak with nice sense, the chaste MIMOSA stands,
> From each rude touch withdraws her timid hands;
> Oft as light clouds o'erpass the Summer-glade,
> Alarm'd she trembles at the moving shade;

And feels, alive through all her tender form,
The whisper'd murmurs of the gathering storm;
Shuts her sweet eye-lids to approaching night;
And hails with freshen'd charms the rising light

are a great joke, playing on its reputation as the 'sensitive plant' and referring in the note to its class, Polygamy, and its defining characteristics. Learning science had rarely been such fun.

This volume had been essentially descriptive, which was why the other – dealing with the principles of natural history – became volume 1. That first volume has for its structure the four ancient elements, fire, earth, water and air. Fire is modernised to include electricity, and earth is the catchword for geology, and magnetism – with an encomium also on Wedgwood and his 'Etruscan' earthenware. The Nile, canals, geysers, fire-fighting, and pumps all feature under 'water', which leads on to sucking children, and an exhortation to mothers to breastfeed. 'Air' naturally refers to Priestley's discoveries, Montgolfier's balloons and the disaster that befell the modern Icarus, Pilatre de Rozier. He it was who set out to fly the Channel in a composite lighter-than-air craft, with one balloon filled with hydrogen and the other with air heated by a brazier: the inevitable spark exploded the hydrogen and dashed the adventurers to pieces on the ground.

Even in this volume there are some evolutionary speculations:[10] 'Perhaps all the supposed monstrous births of Nature are remains of their habits of production in their former less perfect state, or attempts towards greater perfection' or 'do some animals change their forms gradually and become new genera?' and 'I am acquainted with a philosopher . . . who thinks it not impossible, that the first insects were the anthers or stigmas of flowers . . . and that other insects have gradually in long process of time been formed from these'.

Right through there is rollicking verse, making use of gnomes or elves (serious in Shakespeare's *Tempest*, becoming comic here, and nowadays absurd), goddesses, ethereal virgins and nymphs: there is also a discussion of the Eleusinian Mysteries, in connection with the antique Portland Vase which Wedgwood had been trying to copy. What we do not find is serious discussion of God's role in the world; for this Darwin, a child of the Enlightenment, there was no need of that hypothesis. The gnomes and gods were a joke; real explanations would be purely naturalistic. His grandson had more hang-ups.

Erasmus Darwin developed his evolutionary ideas, publishing them in prose in *Zoonomia* (1794–6) and then in verse in his posthumously-published *Temple of Nature* (1803), which came out just after Paley's *Natural Theology*. It belongs to a different world: the authors were contemporaries and these their final works, but Paley was the orthodox clergyman writing lucid prose, and Darwin the poet presenting his secular world-view in playful verse and lengthy endnotes.

The word 'evolution' implied an unfolding, or development according to plan: examples might be the bud evolving into leaves and flowers, or the embryo evolving into the baby animal. The idea of evolution did not have the random, 'higgledy-piggledy' aspect of Charles Darwin's particular theory, to which he preferred to give the name 'development' – this is why the word 'evolution' is absent from the first edition of *The Origin of Species*.[11]

For his grandfather, evolution was a more directed and indeed almost predictable business: and the *Temple of Nature* is concerned with the origin of society, what we would think of as 'social Darwinism' or socio-biology, though his poem is not as ruthless or reductionist as these terms nowadays lead us to expect. The general idea is that the spontaneous generation of living organisms happened when the time was ripe and conditions appropriate; that asexual reproduction followed, leading in its turn to sexual reproduction, and then to society. There is some kind of struggle for existence, in the profusion of seeds from acorns or poppies most of which must die, and among the countless offspring of greenfly and herrings.

While optimism pervades the work, there is a full recognition of pain and death:[12]

> In ocean's pearly haunts, the waves beneath
> Sits the grim monarch of insatiate Death;
> The shark rapacious with descending blow
> Darts on the scaly brood, that swims below;
> The crawling crocodiles, beneath that move,
> Arrest with rising jaw the tribes above;
> With monstrous gape sepulchral whales devour
> Shoals at a gulp, a million in an hour.
> – Air, earth, and ocean, to astonish'd day
> One scene of blood, one mighty tomb display!
> From Hunger's arm the shafts of Death are hurl'd,
> And one great Slaughter-house the warring world!

These are indeed grim thoughts: and show that Erasmus Darwin was nothing like Voltaire's Pangloss,[13] believing naïvely and against all the evidence that everything was for the best in this best of worlds. Nevertheless, the process of evolution had produced us. It had a direction.

We find again in the *Temple of Nature* the ideas that matter is active and endowed with powers, that insects evolved from plants, and that some characteristics have developed in fighting for females. Darwin remarked[14] that

> Many theatric preachers among the Methodists successfully inculcate the fear of death and hell, and live luxuriously on the folly of their hearers: those who suffer under this insanity, are generally most innocent and harmless people, who are then liable to accuse themselves of the greatest imaginary crimes; and have so much intellectual

cowardice, that they dare not reason about those things, which they are directed by their priests to believe. Where this intellectual cowardice is great, the voice of reason is ineffectual.

The only hope was that ridicule might save people from their preachers. Darwin was appalled by the evangelical revival happening in his later years, and by 1800 was out of tune with his times.

He himself met ridicule: his verses had been mocked and parodied by George Canning and Hookham Frere in their periodical, *The Anti-Jacobin*,[15] for Darwin's radical politics (like Priestley's) were seen as unpatriotic when Britain and revolutionary France were at war. His poetry, with its playful Augustan use of classical imagery and elaborate diction, seemed old-fashioned in the new world of Romantic poetry, working with the language of ballads or of ordinary people.

Erasmus Darwin's evolutionary theory offended not merely the revivalists, the religious enthusiasts he targeted, but also men of science reacting against eighteenth-century systems, where everything was accounted for 'in principle' through matter and motion. A firmer empirical base and a tighter fit between theory and observation were demanded in the new world where Francis Bacon was everywhere admired.

We are struck, reading Erasmus Darwin in the light of his grandson's work, by how modern the old man seems: but to contemporaries, he seemed old-fashioned by the end of his life. They had come to expect science to be written in sober prose, its conclusions testable, and its relations to religion less prickly. In England Christianity was seen, in the wake of the bloodthirsty French political experiments, as essential for holding decent society together, rather than as mere priestcraft and superstition from which an enlightened people should be liberated.

Progressive nature (rather than farseeing God) thus lay behind Erasmus Darwin's vision of plants, animals and human society: it was brutal and indifferent to the single life, but it gave a shape and direction to how things turned out, as God had for the Israelites.

Jean Baptiste Lamarck

Progressive nature seems also to lie behind the work of his French contemporary, J.B. Lamarck. He too, by the time he published his evolutionary theories, was seen as out-of-date and confined within an eighteenth-century speculative paradigm,[16] in the same way as Benoit de Maillet[17] in his book of 1748, *Telliamed*, on 'the diminution of the sea'.

Lamarck's greatest work was his book on invertebrates,[18] based on lectures he had given at the great Museum of Natural History in Paris.[19] Insects and worms had been ragbag categories in Linnaeus' system,[20] and Lamarck was one of the pioneers in bringing order: the great majority of animals after all are invertebrates. He emphasised the importance of the backbone as the primary way of dividing the animal kingdom; and in place of the three kingdoms of nature (animal, vegetable and mineral) he saw the importance of distinguishing life – there were two provinces, concerned with the living and the dead, and he was one of those responsible for the term 'biology' as the science concerned with the former, and divided into botany and zoology.

His book contained fold-out tables and descriptions of the classes and genera. Between 1815 and 1822 he expanded this book into a seven-volume study, which at once became a classic of taxonomy. Originally trained in botany, he brought to zoology the art of carefully distinguishing between very similar small organisms, a skill botanists had long had to cultivate. He sought a really natural method, rather than the merely convenient systems that earlier naturalists had adopted.[21]

Informal classification of vertebrates goes back a long way: we distinguish the herring gull, the common gull and the black-headed gull, for example, and all that Linnaeus had to do was to translate these perceptions into Latin (*Larus argentatus*, *Larus canus* and *Larus ridibundus*). They looked different, had different habits, and did not interbreed. It was not difficult to suppose that God had designed each species for its particular niche, having a place for everything and everything in its place. Yet, faced with the enormous numbers of invertebrates – beetles, for example – the biologist, like the layman, finds them hard to distinguish. It may be that God is especially fond of beetles, having created so many sorts and individuals; but it seems more likely that this vast array of similar forms is the result of diversification from far fewer original kinds.

To the invertebrate zoologist, some kind of evolutionary theory was likely to seem more plausible than it would to someone working on birds or mammals; and we may remember here that Charles Darwin's systematic work was done on barnacles. In 1809 Lamarck published his two-volume evolutionary work, *Philosophie Zoologique*.[22] By this time he had a reputation for crankiness: although he had coined the word 'biology' he was not the specialist that men of science in Paris were by then expected to be. He had, in opposition to Lavoisier, devised an alternative way of looking at the facts of chemistry, which to his contemporaries seemed at best eccentric; and between 1800 and 1810 he compiled careful reports on the weather, which seemed tediously empirical, unworthy of the attention of an academician. Moreover, he was at daggers drawn with his fellow-professor of zoology in the museum, Georges Cuvier, whose power and influence were great and increasing.[23]

The *Philosophie Zoologique* was therefore perceived as a throwback to eighteenth-century system building, a whimsical work rather than something serious, and it was ignored. Elderly and distinguished scientists who in their later years embrace offbeat theories find themselves sidelined; it is not uncommon, and so it was with Lamarck. Again, those who make their reputation classifying things are involved in paradox if they then urge that the classes they have painstakingly established are in fact unstable. To the common-sensical, species of animals seem fixed and unchanging – horses and sheep, for example, are different and seem likely to remain so – whereas Lamarck saw species as more like the clouds, rapidly changing their forms.

His rival Cuvier saw the animal kingdom as arranged in four great branches, like a bush: Lamarck adhered more closely to the old idea of a scale, ladder or chain of being, a single line along which organisms could be arranged. Unlike his predecessors, he did not think that this was fixed. His was a dynamical world and animals were moving up the scale with the passage of time: nature was progressive.[24] Whole species could be transformed: his theory was not based upon survival of the fittest individuals. What drove the evolutionary process was the needs of animals, which inclined them to actions, which in turn led to the development of organs. Parts much used increased in size and strength, while those disused contracted and ultimately disappeared.

Lamarck's critics read him as implying that animals wanted their offspring to get on in the world, moving up the scale like good bourgeois citizens; that a tendency to improvement was mysteriously present in nature; and that habits became speedily fixed as characteristics. Cuvier sardonically wrote about how ducks by dint of diving became pikes; how hens searching for food at the water's edge, striving not to get wet, elongated their legs and became storks or herons; and so thousands of species came into being, to provide hard work for those changed by force of habit into scientists.

For Lamarck, nature (rather than every plant, invertebrate or vertebrate) was indeed striving for improvement, with inorganic matter constantly generating the simplest creatures. The environment was not active in shaping animals, but they adapted themselves in response to it – thus determining the exact direction evolution took. There was thus progress and direction, but they were not fully predictable, and indeed had gone differently on different continents. Lamarck believed, like most contemporaries and successors, that acquired characteristics could be inherited; and that this propelled evolutionary change. In the twentieth century, this idea came to be seen as essentially Lamarckian: but in the nineteenth, progress and purpose seemed characteristic (in contrast to natural selection). By 1815, Lamarck had perforce to abandon his simple chain, and adopt two branches for invertebrates (rather than the three Cuvier allocated to them); making his evolutionary scheme less clear.

Cuvier outlived Lamarck, and wrote his obituary for the Academy: it was so slashing that he was asked to tone it down. Debate in France continued, Cuvier and his disciples having on the whole the better of it, notably against Etienne Geoffroy Saint-Hilaire. In Britain, Charles Lyell wrote a chapter against Lamarck in his *Principles of Geology* (1830–33), making him seem so absurd that nobody should take him seriously; and Lyell did not repudiate this until well after his friend and protégé Charles Darwin had published *The Origin of Species* (1859), which made Lamarck suddenly seem a prophet rather than a throwback. This refutation helped to draw attention to Lamarck; and there were some Lamarckians in both France and Britain, seeing directed evolution rather than one-off divine designing as the key to nature's diversity; but until the 1840s they were relatively obscure and isolated from the mainstream of thinking in biology and geology.

Chambers' *Vestiges*

That changed in 1844 with the publication of the anonymous evolutionary book, *Vestiges*,[25] which brought into open discussion the question of how the world had begun and developed. The book was very unusual for a scientific work in being anonymous: reviews were unsigned in most nineteenth-century periodicals, though the French idea that they should be acknowledged slowly spread to other countries: the celebrated Unitarian mathematician Augustus de Morgan has an interesting discussion of the system and its consequences.[26] Much gossip and correspondence among intellectuals concerned the supposed authorship of reviews; but because intellectual worlds were small, and because reviewers were proud of having a polished essay published, the secret was generally soon out. De Morgan describes strategies whereby, without actually telling a lie, a writer could tease a questioner by neither denying nor confirming authorship.

In the case of *Vestiges* the speculation, as Jim Secord shows in his new edition of the book, made the work even more of a sensation; and while many people had guessed who wrote it, the secret was not divulged until after the author's death in 1884. More importantly, *Vestiges* sold 23,350 copies in eleven editions in Britain, with Dutch and German translations, and at least fifteen American editions. It made a big impact, its readers including Abraham Lincoln, Benjamin Disraeli (who was enchanted with it), Alfred Tennyson, Ralph Waldo Emerson, Elizabeth Barrett Browning, Florence Nightingale and George Eliot.

The author was Robert Chambers, who with his brother William had built up a publishing empire in Edinburgh using the new technology of steam presses and cheap paper and contributing much to the 'March of Mind'. He has often been portrayed as an amateur, but that is misleading in the scientific community of the 1840s, when very few could be thought of as 'professional'. Indeed, that word implied to Faraday someone who earned his living doing routine analyses for a fee rather than someone doing 'blue skies' research.[27] Chambers learned his science from the books he was publishing, from background reading and from attending meetings such as those of the British Association for the Advancement of Science. He was up to date, though naturally not everything he picked up was going to stand the test of time – science always has been provisional.

The brothers began *Chambers's Edinburgh Journal* in 1832, and it became an astonishing success among the aspiring middle class: respectable and concerned with improving topics, it was not anti-religious but it was secular – and thus hostile to radical evangelicals. This provoked the minister of the (Presbyterian) Church of Scotland that Robert and his family attended, and he denounced them from the pulpit. They walked out. Anne, his wife, and their children went to the Episcopal Church instead, but Robert was confirmed in anti-clericalism.

Vestiges was denounced by the zealous as atheistic and materialistic, rather as the secular University College, London, had been: but Chambers had taken great trouble in writing it not to provoke evangelical reactions, and to modern eyes the book seems rather over-stuffed with references to God. Despite that, one of the reasons for anonymity was his fear that many subscribers would cancel their orders for Chambers' publications if he were known to be an evolutionist. Many read the book as threatening to religion, or were put off reading it by reviews that raised the bogies of atheism and materialism.

Charles Darwin, who had worked out the basis of his theory of natural selection, was fascinated and appalled by the reactions to *Vestiges*, which he saw as preparing the ground for his work, which was to be based on mastery of enormous masses of data from his own research and that of others. Since 1859 most people have seen *Vestiges* as a forerunner of *The Origin of Species*, which in some sense it was; but that was not how it looked in 1844. And while *The Origin of Species* had limited objectives, concerned with the development of species from a common ancestor, *Vestiges* boldly proclaimed a dynamic worldview. It began with a primeval nebulous 'fire-mist' out of which suns and planets coalesced, it described the spontaneous generation of life and its diversification, and it ended with human society. It was a work of synthesis, and this makes it much closer to our main theme. Chambers' reception from church people and those prominent in the scientific community is still instructive for those building bridges today – the go-between is not much loved.

After half a century of cautious fact- and law-seeking, in the wake of the French Revolution, it seemed to Chambers that the time was ripe for a new synthesis; and he hoped that men of science would welcome his attempt to see the wood rather than just the trees. He drew especially on Babbage's *Bridgewater Treatise* in his discussions of law, on J.P. Nichol's *Architecture of the Heavens* for his astronomy, and on Lambert Quetelet's *Treatise on Man* (published by Chambers) for his statistics.

Nichol was Professor of Practical Astronomy in Glasgow; his book[28] is splendidly illustrated – its second edition (1850) including allegorical pictures by the romantic Edinburgh artist David Scott as well as plates of nebulae – and written in fervid prose well-calculated to get across sublimity. It was Nichol who gave the young William Thomson an impulse towards physics, which ended up with him working on the second law of thermodynamics and the Atlantic cable, becoming the doyen of classical physicists ennobled as Lord Kelvin.

After a complex process of copying his manuscripts and transmitting them through intermediaries, he found a publisher, John Churchill, in London.

Churchill also published *The Lancet* – a medical journal now highly respectable but then deeply critical of the stuffy and stuck-in-the-mud medical establishment – and a strong list of medical and scientific authors, including George Fownes, who had written on chemistry as exemplifying the wisdom and benevolence of God for the Actonian Prize at the Royal Institution, and W.B. Carpenter, a prominent physiologist and Unitarian. Both of these were to assist in updating subsequent editions. *Vestiges* is written accessibly, and carries the reader along with an easy rhetoric, like Paley using familiar analogies to make understanding seem straightforward – though terms like 'fire mist' (for the primeval gas from which everything arose) brought upon him the wrath of pedantic reviewers.

Beginning with the nebular hypothesis, which Chambers believed to be verging upon ascertained truth, he went on to add that bodies were still being formed in space, in the nebulae visible through the largest telescopes. The present was the key to the past. Development, creation, was still going and scientists could study it. The Earth, as one of a 'democracy' of planets, is an indication of what the others are like.

Then follows a gallop through the Earth's strata, emphasising (like Buckland) both the progress we find towards present-day animals and plants, and the changes in conditions upon the globe: land animals and plants could not appear until there was dry land, and as a consequence fresh water. A fundamental plan could be traced through the vertebrates, for example, which varied to suit conditions. This was the key to Chambers' understanding of change.

Aware of Carl von Baer's recent work, including his first observation of the mammalian ovum, Chambers noted how mammalian embryos first look like fish embryos, then like reptile embryos, and only later acquire their distinctive characters. The key to evolutionary change seemed to be that embryos of an existing species remained a little longer in the womb or egg, and developed further. This was a jerky process, worthy of the railway age in which it was conceived: like a railway, embryonic development has only a limited number of branch lines leaving the main line. If an embryo has missed its turning, it must carry on to the next. Individual embryonic development paralleled the evolutionary history of species,[29] or 'ontogeny recapitulates phylogeny' as it was subsequently put. Chambers noted the importance of rudimentary organs, but was unwise in sketching a particular two-stage evolutionary history, from goose to duck-billed platypus to rat.

He faces the issue whether it is degrading to us to think of the Creator as working through evolutionary processes to produce humans; and he concludes it is not. God works through laws, and animals should be seen as parts of a grand plan that only approaches its perfection in ourselves. There is no shame in our genetic connection with them, but rather, good reason to treat them properly.

The normal distribution of Quetelet

Chambers, like Erasmus Darwin, saw no reason to exclude humans from his scheme, and he drew upon the work of Quetelet[30] in order to extend his vision of universal law to our disorderly-seeming realm. Quetelet was employed at the Observatory in Brussels as a statistician. When two people make an astronomical observation (such as the direction of a star or comet) or when one person observes it several times, the results vary slightly. Everybody knew that some observations (and observers) are more accurate than others; but determining the true value was very tricky. People tended to average, leaving out 'poor' values.

Then, at the beginning of the nineteenth century, the great mathematician K.F. Gauss of Göttingen worked out how results are scattered about the true value according to what was called the 'error curve', our Gaussian distribution. Gauss and Laplace were pioneers in bringing certainty and predictability into probability, which had seemed quintessentially chancy – and this quantification of uncertainty into statistical laws was one of the great triumphs of nineteenth-century science, taking it away from the dominance of clockwork causality.

Observatories now needed statisticians and Quetelet was one, in the new little model state, Belgium. He expanded his empire to include human affairs: the error curve became the 'bell curve' on which all sorts of human qualities fall. By the 1830s modern states like Belgium (and even backward Britain) were accumulating information about their citizens to an extent rarely dreamed of before. There were regular censuses, and central recording of births, marriages, deaths, epidemic diseases and serious crimes.

Quetelet found that, whereas people were notoriously (or delightfully) unpredictable even to their families and friends, in the mass humanity was predictable – even (or especially) when it was not law-abiding. Murders, he knew, were done for all sorts of reasons and yet there was a steady murder-rate in Brussels – he chilled the spines of lecture audiences by pointing this out. His work provided the basis for Chambers' deterministic view. It has much in common with that of phrenologists (with whom Chambers was closely associated), for whom our character was determined by the shape of our brains, revealed by the bumps on our skulls. Such ideas may have gone down better in Scotland, with its Calvinist tradition, than in places where free will was more emphasised.

Chambers was fascinated with Swainson's 'Quinarian' system of classifying animals,[31] in which there was a pattern of circles into which everything fell, giving a tidy and satisfying shape to nature rather than the disorderly and unpruned trees and bushes of Lamarck and Cuvier. He also saw no real reason for believing that we were necessarily the terminus of the evolutionary process. Mothers might in the future give birth to super people, no longer half-akin to brute. This was an idea that helped Tennyson to come to terms with the death just after graduation of his friend (and his sister's fiancé) Arthur Hallam, who had seemed destined for a great career – he had perhaps been born before his time.

Chambers was disappointed to find his book being denounced on all sides – by the clergy as being atheistic and among biologists as inaccurate – though the hostile reviews did draw attention to the book and promoted its sale. From our point of view, it seems there were few who could imagine God working through laws over a very long period rather than intervening supernaturally:[32] 'Mr Vestiges' looked at best like a Deist, not a proper Christian. Sedgwick saw *Vestiges* as unmanly, probably written by a woman (no lady) to mislead others of her sex into materialism. Others at Cambridge, notably Whewell and Herschel, were alarmed because the basis of their compromise with religion was that science simply sought to understand the world as a going concern. It was not to be concerned with origins or with the radically new: these were the domain of religion, no place for the speculations of scientists.

For Chambers, God was working out His purposes in ways that his critics did not want to recognise, because they felt this moved religion from the middle of the picture to a small place in the background. In the middle years of the century, it seemed best to the respectable to avoid big questions to do with evolution. It seemed the same to the agnostic Huxley who – because he did not want to deal with those barren virgins, final causes, and raise unanswerable questions about ultimate purposes – had to be content to describe and classify. *Vestiges* had an immense influence, but largely among lay people, keen to see a direction in their world, rather than among clergy or clerisy.

Darwin's degenerate barnacles

Charles Darwin was uneasy about progress, because he saw the world as governed by natural selection, which was about filling niches rather than going upwards. He disliked talk of higher and lower organisms; his close study of barnacles had indicated how animals can go down in the world. The offspring of barnacles are like little shrimps and, though barnacles look rather like limpets, they belong to the same group as lobsters. The naval surgeon J.V. Thompson stationed in Cork, Ireland, had established this in the years 1828–34;[33] but many had found it hard to believe. Darwin's formidable volumes on living and fossil barnacles confirmed their relationships,[34] and among the numerous species he described some where the male had degenerated even further, becoming no more than a stomach and sex organs living within the shell of the female.

This was far from Victorian Values, but degeneration and throwbacks might equally happen among humans. Darwin's poor health led him to worry that it might be inherited by his children, and others began to fret publicly about those of low intelligence and criminal propensities reproducing themselves.

Whereas Lamarck and Chambers had been optimistic, seeing nature producing better and better creatures as time went by, Darwin saw that not all the animals and plants of remote epochs had perished.

Some familiar species closely resembled those existing tens of millions of years ago: a good design, and an environment which has not changed too radically, could lead to a very long run. This undermined Lyell's suggestion that species, like individuals, might be young, and then age and die: instead, we might go on and on. It was the thought of degeneration, though, that caught the attention of many in the later nineteenth century, when *fin-de-siècle* gloom competed with confidence in progress for the public mind.

The gloom was fuelled by recent work in the physical sciences. Kelvin was one of the pioneers of thermodynamics, which in mid-century transformed our understanding of the world.[35] Conservation of energy was the key to understanding changes right across the sciences: and its acceptance led to the union of electricity, magnetism, mechanics, optics and other branches of knowledge into classical physics.

Chemistry was dethroned as the fundamental science, and threatened with subordination: the new branch, physical chemistry, was a sign that much of the science might be reduced to physics.[36] And then the second law of thermodynamics gave a direction to the world: it was running down.[37] Energy was becoming less and less available: eventually everything hot would cool down, and in the universally tepid world everything would come to an end in what was called 'heat-death'.

Thomson worked out the age of the Sun – on the assumption that it was made of the best Newcastle coal, additionally fuelled by meteors and by gravitational collapse – and the age of the Earth, which he took to be a slowly cooling sphere. Both lines of reasoning converged, which is always encouraging for a scientist: and gave an age of about 50 million years, or 100 million at most. He reckoned that both had much the same life left in them: they were middle-aged. In a few tens of millions of years, then, the Sun would have become a big cool red star and the Earth would be incapable of supporting life. We lived in a world that was perhaps past its peak – that, like us, had a beginning and would have an end.

Thomson's figures did not appear to allow enough time for natural selection to have produced all the variety the world displays, and so they seemed to be evidence for a more directed or jerky form of evolution. He was sure that geologists must take their data from physicists, working in the fundamental science rather than in some more descriptive and less austere discipline. T.H. Huxley remonstrated, but largely in vain – though he was vindicated in the twentieth century, when the discovery of radioactivity indicated that Thomson's figures were far too low. But the projection into the future was more significant, and seemed a reason for nihilism: all human striving and building would come to nothing in the long run, and that was depressing even if there were several million years to go.

The Time Machine: Earth

H.G. Wells had been a student of Huxley's, attending one of his courses in South Kensington and enormously admiring him, and was well aware of current science. His novel *The Time Machine* (published in 1895, the year Huxley died) explores not the years that have passed but those that are to come.[38] This avoided philosophical problems about the time traveller affecting the past, and therefore the present; and also allowed exploration of a world in decline and old age. When the traveller first stops, he is in a world where the decaying ruins of London (including those of a great museum) are to be seen, and the human race has evolved (in fact degenerated) into two species filling different niches. One (Eloi) is descended from the upper classes of Victorian society, pretty, feckless and living on the surface; the other (Morlocks) from the working class, brutal, living underground and coming up at night to prey upon their ineffectual congeners. Evolution, in truly Darwinian style, has not implied progress towards perfection or a higher state of being: it has been change for the worse.

This dystopia is also a commentary upon Victorian Values and the inequalities and class distinctions of Wells' time: this is a powerful fable. His second stop is much further on, and by now the Sun is vast, red and relatively cool; all vertebrate life seems to have disappeared, and in the gloom of a chilly dawn he is beset by some enormous crab-like creature, from which he escapes with difficulty as he had from the brutish Morlocks:[39]

> I cannot convey the sense of abominable desolation that hung over the world. The red eastern sky, the northward blackness, the salt Dead Sea, the stony beach crawling with these foul, slow-stirring monsters, the uniform poisonous-looking green of the lichenous plants, the thin air that hurts one's lungs; all contributed to an appalling effect.

A hundred years earlier, Erasmus Darwin had seen progress and rejoiced in it: though he had no place for a personal God, nature in full development could be source of delight and spiritual satisfaction.

Charles Darwin, coming to terms with nature red in tooth and claw as revealed in fossils and in the countryside, with the death of his daughter Annie, and with his own poor health, boldly faced an open-ended evolutionary process that depended upon natural selection. Wells, exploring this world-view, saw a bloody past and a future that promised no better, and would ultimately and inevitably be much worse. There was no order and purpose to be seen. Mind in despair reached the end of its tether.

We should not suppose that for everyone the 1890s were either 'naughty' or gloomy: labels may draw attention to some features of decades but that is all. My granny was a teenager in that decade, and it was a great lesson in history when she told me that it wasn't like that where she was. With Darwin dead and buried, Darwinism in the strict sense was in eclipse.[40] Evolution was not, for most people a problem: in place of Wells' bleak vision, they held on to the idea that there was direction, and that Darwin's *The Descent of Man* (1871) could be better seen through the earnest eyes of Henry Drummond as *The Ascent of Man* (1894).[41] Egoism and altruism were jointly responsible for evolution, and the past provided us with a guide as to what was possible and best: 'the great Mother in setting their difficult task to her later children provides them with one superb part finished to show the pattern'.

Drummond's liberal view of religion enabled him to fit Christianity into the evolving pattern, and also at the end to look forward to Higher Evolution and the Higher Kingdom founded upon altruism, which after all is another name for charity, philanthropy and love. No doubt those who enjoy good health and a satisfactory income find it easiest to dilate upon the goodness and wisdom of God, and at the end of the nineteenth century there were many (and many more than ever before) in this position in Europe and the USA. They combined optimism and theism with either early versions of the welfare state, as in Bismarck's Germany, or with social Darwinism and eugenics – sometimes both.

The First World War was very shattering, even to those who believed that it would be the war to end war; and easy optimism about progress could not so easily be maintained after it. A new mechanism for evolution, random genetic mutation, was carrying all before it, and natural selection seemed to occupy a very subsidiary role. The Cambridge theologian Charles Raven sought a new natural theology based on non-Darwinian evolution[42] and stout opposition to mechanistic biology, combined with liberal Anglicanism and a firm belief in the reality and importance of the material world in God's plan. But by the 1930s the neo-Darwinian synthesis was emerging, and everywhere mechanists and believers in natural selection were triumphing: Raven found that biologists failed to respond to his call. Within his church, he also found himself marginalised as Karl Barth's neo-Orthodoxy prevailed, and natural theology seemed to most clergy a hopeless cause. Again, the honest broker was unwelcome, and Raven never became the prophet he had hoped to be.

The question of purpose, of intelligent design, is still alive despite all that has happened since Paley, Erasmus Darwin and Lamarck. Altruism remains something to be explained away for out-and-out materialists; purpose, or even function, is an idea to be treated with the greatest caution – though these concepts are often invoked in making sense of new discoveries about the natural world, and anthropomorphism (of a Darwinian flavour) is rife.

The latter is used to ascribe quasi-human capacities to animals in order to work back to an evolutionary psychology where human behaviour is explained in terms of animal ancestry. This may sound more like sermonising than science: and it is lay sermonising to which we next turn – the preaching of science in secular cathedrals, which in the form of public lecturing, became prominent from the late eighteenth century.

CHAPTER 8

Lay sermons

'READ UNTIL YOU HEAR THE voices' is the advice given to those trying to get into the past. Trying to hear voices is tantalising, until we reach the epoch of Edison and can listen to scratchy recordings of Florence Nightingale or Lord Kelvin, and yet the spoken word – the word *par excellence* – is and was of critical importance, in science as in religion. Lectures were the best way to get science across, better than even the most beguiling book, for experiments could demonstrate the message. Nowadays, lectures with a popular slant cut no ice in scientific ranking: only papers read at learned society meetings or conferences feature in scientists' appraisals of their peers, or in rating university departments. The same is true of popular books and even textbooks.

It was not always so: in the nineteenth century some of the best popularising and reflecting on science, its goals and methods, was done by eminent practitioners in the course of duty, and they were given due credit. Going on a voyage of discovery was one way of rising to the top in science,[1] as for Charles Darwin, Banks, Huxley, Edward Sabine and Joseph Hooker; but becoming a famous lecturer worked for Davy, Faraday, Huxley again,[2] Tyndall and Gosse.[3] The relatively unspecialised lecturer was then also expected to reflect upon the science, as a natural philosopher. Such reflection very frequently included references to God's wisdom and benevolence in a manner that would now seem rather startling and eccentric.[4] The scientific establishment needed its great preachers just as the churches did, to get the message across to outsiders and to stimulate the faithful. Sermons are very different from formal, careful expositions of doctrine: they are personal.

Lay preachers were an invention of John Wesley's (or of earlier dissenters), but the more respectable churches allowed only the clergy to enter the pulpit; while to have women holding forth was a sign of a conventicle beyond the pale.[5] Parsons were accustomed to preach sometimes for an hour; and were immune from interruption or contradiction. In the 1860s my grandfather, a parson's son then aged about five, preached a sermon to his brothers and sisters: 'Ye are fools, the lot of ye'. This was unwise. They were bigger, they duly thumped him, and when grown up he found another vocation. All churchgoers must at some time have endured a sermon on the same lines (at much greater length) and boiled inwardly – but really successful preaching requires respect for one's hearers.

A poet puts science in its place

S.T. Coleridge the poet became at Davy's invitation a familiar lecturer on literature and philosophy in London,[6] and gave the title *Lay Sermons* to some of his writings about religion and its place in the national life in 1815. These (though never actually spoken) were like the real sermons he had twenty years earlier preached in Unitarian chapels, until released from this bondage by Thomas Wedgwood's generous offer of £150 per year to write poetry and philosophy.

The 'sermons' begin and end with a verse from the Bible. The nub of the first one, 'The statesman's manual', is: 'WE (that is, the human race) LIVE BY FAITH' in the broadest sense, and should live by biblical faith and morals in a world of overweening philosophers and scientists. France had shown the danger of allowing a presumptuous and irreligious philosophy to predominate, and Coleridge pointed to the danger from[7]

> The extreme over-rating of the knowledge and power given by the improvements in the arts and sciences, especially those of astronomy, mechanics, and a wonder-working chemistry; ... an assumption of prophetic power, and the general conceit that states and governments might be and ought to be constructed as machines, every movement of which might be foreseen and taken into previous calculation; ... the consequent multitude of plans and constitutions, planners and constitution-makers

– and the remorseless arrogance with which they pursued their plans and whim-whams. Coleridge feared that the sermon might seem 'the overflow of an earnest mind rather than an orderly premeditated composition'; but his ideas proved extremely important to a generation uneasy about innovation, as we eyeing recent social experiments should be.

The first sermon was addressed to the learned, the second to the middle classes; but the messages are much the same for a society emerging from decades of war, facing economic depression and demands for political reform. Here too, Coleridge was unhappy about science – knowledge rather than wisdom – and even about the high reputation of his friend Davy:[8]

> If Plato himself were to return and renew his sublime lucubrations in the metropolis of Great Britain, a handicraftsman, from a laboratory, who had just succeeded in disoxydating an Earth, would be thought far the more respectable, nay, the more illustrious person of the two.

The 'handicraftsman' alluded here must be Davy, who had isolated new metals by electrolysis in the laboratory at the Royal Institution directly beneath the great lecture theatre where Coleridge was speaking. Since Davy had rescued Coleridge from despair by arranging for him to lecture, right above the laboratory where calcium and other 'alkaline earth' metals had been discovered, there is an element here of biting the hand that fed him.[9]

Whewell in 1833 coined the word 'scientist' in response to Coleridge's criticism of those working in empirical science appropriating the term 'philosopher', even when (unlike Davy) they might be uninterested in general principles or world-views. It would be wrong, though, to suppose that the sciences were Coleridge's target, or that he saw them as a threat to true religion. The things he detested were ruthless and oppressive Malthusian 'political economy', radical politics on the French revolutionary model, and the utilitarian ethics of Paley and Bentham – all claiming some scientific authority. At the same time he deplored what he saw as the failure of the Church of England in its social and educational role:[10]

> What are all these Mechanics Institutions, Societies for spreading Knowledge, &c. but so many confessions of the necessity and the absence of a National Church?

We shall explore in another chapter Coleridge's vision of a 'clerisy' that would augment or replace the clergy and Huxley's idea of a 'Church Scientific'; but Coleridge has led us to the institutions within which science was preached. We must return to our project of hearing voices.

Huxley: chalk and talk

Let us try listening to Huxley, the young ship's surgeon who on his return from his voyage on HMS *Rattlesnake* found himself famous and respected as a scientist (elected FRS) but unable to find a job. He decided to leave the navy and devote himself to science, particularly to writing up his discoveries among marine invertebrates; but, in order to live, he turned perforce to journalism and lecturing. All his life he was short of money, but this hard training in putting himself across paid off: he became a sharp, incisive and direct writer, and a great speaker. Here he is, focusing upon a piece of chalk in a lecture to working men:[11]

> A great chapter in the history of the world is written in the chalk. Few passages in the history of man can be supported by such an overwhelming mass of direct and indirect evidence as that which testifies to the truth of the fragment of the history of the globe, which I hope to enable you to read, with your own eyes, tonight. Let me add, that few chapters of human history have a more profound significance for

ourselves. I weigh my words well when I assert, that the man who should know the true history of the bit of chalk which every carpenter carries about in his breeches-pocket, though ignorant of all other history, is likely, if he will think his knowledge out to its ultimate results, to have a truer, and therefore a better, conception of this wonderful universe, and man's relation to it, then the most learned student who is deep-read in the records of humanity and ignorant of those of Nature.

It was through Huxley's admirer H.G. Wells that schoolchildren's history now includes palaeontology, in the belief that dinosaurs are as relevant to us as the feudal system. The rhetoric is splendid; but Huxley like other good lecturers never read his script, and probably spoke with no notes or very minimal ones, expertly drawing pictures on the blackboard to illustrate his points. His performances were a spectacle.[12] Written and spoken English are different enough for us to be sure that the listeners heard something distinct from what we have in front of us – but the tone and content are there, even if Huxley in full flood eludes us.

The move from the apparently banal or ridiculous to the sublime may remind us of Bishop Berkeley getting rapidly from tar water to the Holy Trinity.[13] Huxley's wished to make science a part of general culture, with an input into its values and beliefs. He strongly believed that everyone ought to know some science – truth would set us free – and be familiar with its methods of finding truth. His eminence in research ensured that he was not seen as a mere populariser, a Mr Vestiges:[14] and in his time the distinction between the full-time, active scientist and the amateur was beginning to become clear. We should be careful of setting 'professional' against 'amateur' for the terms were not opposites then. Huxley's friend and ally Joseph Hooker[15] (like Faraday) distinguished his 'professional' activities – the chores connected with his post as director at Kew – from his 'philosophical' ones, his botanical research.

Where Huxley was new was not in bringing religion into scientific lectures, but in refusing to yield authority to religion in scientific matters. He could get away with this without the taint of 'irreligion' because, in contrast to a previous generation, he lived a respectable life – as Nietzsche amusedly noted of many English doubters.[16] It was true that Mary Ann Evans had lost her faith translating German theology, had moved in with a married man and wrote novels under the name George Eliot;[17] but nobody could convincingly say (though some tried)[18] that Huxley was promoting immorality.

He was particularly keen to distance himself from the positivism of Auguste Comte (with whom critics kept trying to associate him) because that seemed to have undertones of French immorality, metaphysics and pseudo-religion. Huxley proved empirically that religion and ethics were not necessarily entwined. He and Hooker particularly disliked the influence of the clergy –

their grasp on the educational system, their clannishness and adherence to the party line – and felt, in the last resort, they could not be trusted. The man of science had to be open-minded, agnostic, without commitments or prejudices, and thus inevitably critical of dogma.

Huxley's background had been not unlike that of Davy but, where Davy took advantage of the patronage system (notably through Banks), Huxley fifty years later rejected it. Indeed, a refusal to be patronised can be seen as a key to his life.[19] Davy escaped the drudgery of 'professional' duties by marrying a wealthy widow; Huxley's beloved Henrietta, met in Australia, brought him energy, devotion and children but no money. Where these two masters of scientific oratory were at one was in their vision of a society that took full advantage of science, and in their public view of the nature of science.

Davy and the wars against the French

Davy's inaugural lecture of 1802 at the Royal Institution had ended with this familiar flourish:[20]

> In this view, we do not look to distant ages, or amuse ourselves with brilliant though delusive dreams concerning the infinite improveability of man, the annihilation of labour, disease, and even death, but we reason by analogy from simple facts, we consider only a state of human progression arising out of its present condition, – we look for a time that we may reasonably expect – FOR A BRIGHT DAY, OF WHICH WE ALREADY BEHOLD THE DAWN.

Later, in his *Consolations in Travel* (much of it recycled from Royal Institution lectures) Davy wrote[21] that 'science is nothing more than the refinement of common sense making use of facts already known to acquire new facts'. Huxley took up this phrase in his celebrated definition:[22] 'Science is, I believe, nothing but *trained and organised common sense.*' The two adjectives are very important, as we shall see; but crucial here is that these two great lecturers were out to demystify science.

Davy and Huxley knew that science could not be carried very far by those of ordinary intellect,[23] but they did not seek (as some popularisers do) to blind or dazzle their hearers with science: its methods and conclusions were a refinement only of what everyone does all the time. Lecture audiences could feel themselves drawn in to the process of discovery and explanation, the very life of science, illusory though this might be – sadly, and the lecturers knew it,[24] Davy's plutocrats or aristocrats, and their wives and daughters, were not going to become scientists any more than Huxley's working men. But they could empathise. Science was not some arcane mystery, understood only by mandarin geniuses in ivory towers propped up by an uncomprehending public: it was open to all, like Christianity. But it needed to be got across.

Generals in the army of science

Aggression can be a wonderful fuel for a public speaker. Prophets and preachers have seldom had a wholly benign message: they denounce sins and probe their audience into discomfort. Because war with France was a feature of most of Davy's working life, the aggression that fuelled his ambition could be directed at his country's enemies. His chemical discoveries, which set straight French mistakes, were so many victories in the wars. In a lecture demonstrating the falsity of Lavoisier's idea that all acids contained oxygen because there is none in the acid from sea salt, he exulted that:[25]

> No part of modern chemistry has been considered as so firmly established, or so happily elucidated; but we shall find that it is entirely false – the baseless fabric of a vision . . . The confidence of the French inquirers closed for nearly a third of a century this noble path of investigation, which I am convinced will lead to many results of much more importance than those which I have endeavoured to exhibit to you. Nothing is so fatal to the progress of the human mind as to suppose that our views of science are ultimate.

The enemy of truth was C.L. Berthollet, most exalted of French chemists, whose dogmatism could be duly deplored by Davy's excited audience (including the young Faraday taking notes to bind up and present to the great man, with the request for a job).

Davy's tone was suitably belligerent, and a quotation from Shakespeare[26] appropriate. Not long afterwards Davy (now Sir Humphry) visited France to collect a prize from the Academy of Sciences, taking Faraday with him. They played the role of Coriolanus, fighting for their country in the very metropolis of the enemy, as they raced with Gay-Lussac to elucidate the nature of iodine, which Davy saw as an analogue of the element he had recently named chlorine.

Preachers may also very often have risen to prominence from a modest background: the church has been a wonderful vehicle for social mobility. By the nineteenth science was beginning to play the same role. In a ploy to dispel Humean doubts about the Gospels or the very existence of Jesus, the logician Richard Whately (later Archbishop of Dublin) wrote a splendid skit[27] in which the life of Napoleon (still alive, but in exile) was presented as a fiction, an excuse for raising taxes by spinning a yarn about a poor boy from Corsica who in classic fairy-tale manner becomes Emperor, is crowned by the Pope, conquers the world and marries the Princess. Davy, Faraday and Huxley were less spectacular examples of upward mobility, but impressive nonetheless. Faraday was a saintly man, but the other two were more ambitious, ready to play their part in organising science but expecting due recognition.

Davy had enemies nearer home when he became president of the Royal

Society in 1820. He found himself as a general in the army of science[28] directing his social superiors in a very class-conscious society. Huxley, born in 1825, by contrast belonged to a generation which lived through profound peace: British wars did happen, in the Crimea, Bengal and Africa, but they were sideshows; and the wars in the USA, Germany and France did not threaten independent Britain as Napoleon had. Huxley's enemies were therefore not national, but local: notably Richard Owen, who had risen like Davy through patronage to be the country's leading fossils expert. Anyone else prominent among the powers-that-be in church and state – what we call the 'establishment' – was also fair game to Huxley, including the clergy and genial Oxford or Cambridge men who were not really deeply concerned with knowledge and truth.

Huxley's ferocity became notorious, so that in 1864 a favourable review of his writings on comparative anatomy, especially skulls, noted that:[29]

> As far as the public are concerned, they either take it as a matter of course that Professor Owen will be attacked whenever Professor Huxley speaks or writes; or they crowd to the lecture hall with the same feelings as they would go to witness a prize fight; all that we can say is, that it imparts to the non-scientific world a false estimate of the spirit which exists among scientific men, a very false estimate indeed, and what chiefly concerns us as reviewers is that it does great permanent injury and reduces the value of an author's works, for it is difficult to accredit a writer with strict impartiality, who cannot exercise a little control over his feelings.

Impartiality should have been one of the virtues of agnosticism, but Huxley's formidable style had more affinity with the law court than with the calm temple of Minerva.

Huxley and the war against ignorance

We can see Huxley making war in his Royal Institution discourse of 10 February 1860. These discourses were organised by Faraday for Friday evenings, and they still continue: distinguished scientists were invited to review their own work and the state of their discipline, taking advantage of the equipment of the lecture theatre to demonstrate experiments or specimens – they were not expected to be polemical, and they were to aim at an educated, polished but unspecialised audience. Huxley had held a part-time post there, as Fullerian Professor, since 1855, and gave a discourse each year, as did Owen, Tyndall and Faraday. Other lecturers included William Thomson, Darwin's Captain (now Admiral) FitzRoy, Lyell and John Ruskin.

The Origin of Species had been published late in 1859 and reviewed by Huxley in *The Times* (England's most prestigious newspaper) the day after Christmas.

Now he was to hold forth on 'Species and races and their origin': Darwin had been consulted, and different varieties of pigeons bred by fanciers were to be on display. The lecture fell into two parts: the first was taken up with Darwin's book and how far 'domestic selection', the breeding of pigeons and livestock, justified the notion that nature 'must tend to cherish those varieties which are better fitted to work harmoniously with the conditions she offers, and to destroy the rest'.[30]

For the second half of his lecture, Huxley hit a problem: Darwin's theory of natural selection did not measure up to the stringent conditions for real science that Huxley had earlier laid down, mentioning Tyndall's work on glaciers where imagination was controlled and tested by experiment.[31] It must therefore count as only a hypothesis. There were no experiments to demonstrate the production of species, or proof that such a process ever happened in nature.

These were criticisms that many scientists made at the time. Indeed, it is likely that more clergymen went along with Darwin's ideas – much of them, at least – than did people holding scientific posts.[32] His theory was neither inductive in the Baconian manner, generalising from a number of cases in which one species had changed into another (for such a thing has never been witnessed), nor deductive, with definite and testable consequences open to experiment and observation. It was indeed a 'common-sense' extrapolation in a Britain preoccupied by the industrial revolution and Malthusian demography, and that is probably why there were several Britons who proposed theories more or less like Darwin's, but nobody from Continental Europe.[33]

Huxley, like many contemporaries, never managed to find natural selection easy to accept as a major mechanism for evolution: the very different-looking varieties of pigeon with which he illustrated his lecture were after all members of the same species, all bred from the rock-dove and inter-fertile like different breeds of dogs. Darwin had written that, if these pigeons had been found flying around in some remote region, they would have been classified as distinct species; but that did not satisfy Huxley. He remained uneasy about Darwin's 'hypothesis', rather as the inquisitors had been about Galileo's, though Huxley would not have relished that comparison.

He therefore turned to a good old rant for the second half of his discourse. He attacked 'unscientific' objections to Darwin as foolish, frightened and illogical:[34]

> The man of science is the sworn interpreter of nature in the high court of reason. But of what avail is his honest speech, if ignorance is the assessor of the judge, and prejudice foreman of the jury? . . . And there is a wonderful tenacity of life about this sort of opposition to physical science. Crushed and maimed in every battle, it yet never seems to be slain; and after a hundred defeats it is at this day as rampant, though happily not as mischievous, as in the time of Galileo.

There were unfortunately 'foolish meddlers' who thought they did 'the Almighty a service by preventing a thorough study of his works'; who were ashamed of parts of the glorious fabric of the world.

Huxley in his peroration moved like Davy into patriotic mode, no doubt appropriate as France under Napoleon III seemed to be looking for new worlds to conquer and the United States drifted into civil war. He noted an intellectual revolution happening:

> The part which England may play in the battle is a grand and a noble one. She may prove to the world, that for one people, at any rate, despotism and demagogy are not the necessary alternatives of government; that freedom and order are not incompatible; that reverence is the handmaid of knowledge; that free discussion is the life of truth, and of true unity in a nation. Will England play this part? That depends upon how you, the public, deal with science. Cherish her, venerate her, follow her methods faithfully and implicitly in their application to all branches of human thought; and the future of this people will be greater than the past. Listen to those who would silence and crush her, and I fear our children will see the glory of England vanishing like Arthur in the mist; they too will cry too late the woeful cry of Guinever:
>
>> It was my duty to have loved the highest;
>> It surely was my profit, had I known
>> It would have been my pleasure, had I seen.

The stanza from Tennyson[35] makes a wonderful ending (pulpit eloquence) to what must have been a sensational lecture, though not what Darwin had hoped for and expected.

The rhetoric of warfare has its problems: where Darwin would have liked to win people over so that by slow degrees they slid down the slippery slope, Huxley instead polarised people by making them take sides. His rhetoric, with his claim to be the honest champion of truth and science against ignorance and prejudice, was not new; but it was new to his liberally educated hearers at the Royal Institution. During that summer, Huxley debated evolution against Samuel Wilberforce in Oxford. This poorly-reported event, where the most telling blows against Wilberforce probably came from Hooker rather than Huxley, was inaugurated by a lengthy speech from the visiting American, John William Draper; and in 1874 his *History of the Conflict between Religion and Science*[36] implied that a state of warfare and aggression was the natural relationship.

Draper's development of Huxley's position – science vanquishing unsophisticated religion – has been astonishingly successful as a 'grand narrative' of the nineteenth and twentieth centuries. Twenty-one years later at the Royal Institution, Huxley again used the language of war but now of triumph: 'I think it is to the credit of our age that the war was not fiercer, and that the more bitter and unscrupulous forms of opposition died away as soon as they did'. Huxley was also able to assert, following his own work (and Owen's!) on *Archaeopteryx* and the reptile–bird relationship, that 'evolution is no longer a speculation, but a historical fact': Darwin had lived long enough 'to see the stone the builders rejected become the head-stone of the corner'.[37]

Fitting in a biblical text in this context made the peroration a *tour de force*. But while in Huxley's time the power of the churches was nothing like it had been in Galileo's, it was still something: whereas Huxley's successors in our day, like Richard Dawkins, seem by contrast to be tilting at windmills. We have more potent absurdities than religion opposed to reason and science.

Uncommon sense

Faced with a prosperous and well-educated audience, Huxley could tease them; he treated working men with more respect. He saw himself as a plebeian, with a duty to help those from the working classes; and his was a generation anxious for self-improvement, practising self-help. Invective and polemic were out of place; it was more appropriate (and important) to be earnest. We return to that piece of chalk, the topic of his lecture in 1868 at the British Association meeting in Norwich, and his message that science was trained and organised common sense.

It was true that Huxley had used his medical training to recognise that the hips of some dinosaurs resembled those of ostriches, and they therefore roamed around on two legs rather than slithering like alligators – but, even if ordinary people could not have made that inference, they could follow it afterwards. Again, without organisation, without the numerous bones from fossil and living species collected and brought together in museums,[38] comparisons would have been impossible. But after that it was common sense, and the man of science and the men of sense were, as Huxley put it, more or less on a level.

His strategy here was to make hearers realise that they did science all the time, just as Molière's M. Jourdain spoke prose all the time. When the evidence was clearly presented, an audience of working men would, like an open-minded jury, come to the right conclusion. Lawyer-like he drew them along, 'the great and beloved teacher, the unequalled orator, the brilliant essayist, the unconquerable champion and literary swordsman'.[39]

Having presented the evidence, notably on different kinds of crocodiles, Huxley summed up with a rhetorical flourish. The working men were invited to[40]

> Choose your hypothesis; I have chosen mine. I can find no warranty for believing in the distinct creation of a score of successive species of crocodiles in the course of countless ages of time. Science gives no countenance to such a wild fancy; nor can the perverse ingenuity of a

commentator pretend to discover this sense, in the simple words with which the writer of Genesis records the proceedings of the fifth and sixth days of the Creation.

For Huxley, the operation of natural causes underlay everything that we see. His conclusion was a magnificent example of using illuminating metaphors from science:[41]

> A small beginning has led us to a great ending. If I were to put the bit of chalk with which we started into the hot but obscure flame of burning hydrogen, it would presently shine like the sun. It seems to me that this physical metamorphosis is no false image of what has been the result of our subjecting it to a jet of fervent, though nowise brilliant, thought tonight. It has become luminous, and its clear rays, penetrating the abyss of the remote past, have brought within our ken some stages of the evolution of the earth.

His congregation would have been edified without being mystified or dazzled by such limelight.

Unlike the students who had examinations to pass, or the Royal Institution grandees whom it was fun to shock, the working men were very unlikely to practise science. Huxley was casting his bread upon the waters: working men would 'do science' only in their day-to-day, common-sense reasoning. They could not be trained and organised into scientific research, but some science would enable them to think clearly, detect false logic and become more responsible citizens. It also opened sublime vistas. It was an important part of the education of everyman.

For Huxley's students, getting a course on the crayfish, it was different – their lectures were not lay sermons. University lecturers, like 'selfish genes', long to replicate themselves, and science courses are always based on the assumption that the hearers will become active scientists. Huxley was indeed 'father in science' to the most prominent physiologists and zoologists of the next generation, but they were puzzled by his career. However effective a lecture or sermon may be, the world will little note the details nor long remember them.

Huxley's reputation was in his time enhanced by his lecturing; but to his greatest pupil, Michael Foster, it seemed self-sacrificing:[42] 'to a large extent [he] deserted scientific research and forsook the joys which it might bring to himself, in order that he might secure for others that full freedom of inquiry which is the necessary condition for the advance of natural knowledge'. We shall not be tempted to see Huxley as a Bodhisattva, because his bravura performances brought him personal satisfaction and the very public career that he wanted.

Huxley was the most eminent and successful of many who, by the 1870s, were keen to attack organised religion and replace it by something more dynamic and modern.[43] With some of these, like the disciples of Auguste Comte, Huxley wanted little to do – he disliked positivism as combining scientific and religious doctrine ('Catholicism minus Christianity')[44] – and he was also very chary of whatever smacked of immorality. In 1887 he was president of the Sunday Lecture Society, which met at four o'clock each Sunday, between November and May, at St George's Hall in Langham Place, in London's West End, to hear talks directed at ignorance and superstition. The lectures were published, price 3d: one could become a member of the society for £1 per annum. Tickets for individual lectures cost a penny, sixpence or a shilling (depending where one sat), but tickets for the series of twenty-four lectures cost only 5/6 for the shilling seats, or 2/- for the sixpenny seats, and were thus a very good bargain. We should remember that, at this time, pews in most churches were rented for the year: they were free only at the back or in the gallery.

The Sunday Lecture was not altogether unlike churchgoing. The themes were often scientific: the movements of plants, instincts, and the nature of atoms were among the topics. Others included the American experience of co-education, the lives of Rousseau and Voltaire, matter and soul, and the analysis of curious religious beliefs. The society[45] aimed to cover science (physical, intellectual and moral) and also history, literature and art 'especially in their bearing upon the improvement and social well-being of mankind'. Organised irreligion, almost as a mirror image of religion, had arrived; and in 'ethical societies' – which closely paralleled Dissenting chapels – high-minded agnostics and atheists met for lectures and earnest discussion.[46]

Cleanliness is next to Godliness

It is one thing to have edifying lectures or sermons, and quite another to cope with the disasters of life. There had been illness called cholera before, but nothing like the 'Asiatic cholera' that arrived in Europe in 1830.[47] This new kind was terrible: people who had been healthy in the morning died in agony, turning a horrible bluish colour, by afternoon, following catastrophic diarrhoea. The plague had spread irresistibly from India across Asia to Europe; and hopes that Britain might be spared by quarantining ships were dashed, when it broke out in Sunderland and spread through the country. People had lived with death rates, especially among babies and young children, that would appal us; and they took for granted that many young adults would be carried off by tuberculosis. But a new disease was terrifying, and to many it seemed that God must be wrathful.

Ever since Jeremiah, catastrophe has been perceived as judgment: just as God punished the Egyptians for enslaving the Israelites, so He would punish other wicked nations with plagues. There was no doubt that Britain was (like others) arrogant and sinful, though it might not be clear exactly which sins had especially affronted God. The only escape was to do as the people of Nineveh had done when alerted (belatedly, by Jonah after his adventure with the whale), and humbly repent with fasting and prayer. Days of national humiliation and prayer were duly called, in an unusually ecumenical spirit; and clerical sermons brought home to people their collective and individual sins.

In the twentieth century, this acceptance of collective sin and judgment was largely lost: indeed it had been cast aside by the highly personal and individual evangelical Christianity that flourished in the late nineteenth century, when the clergy called for individual repentance. But even in the 1830s, social and collective responsibility were starting to be seen as sanitary rather than spiritual. There were calls from medical and social reformers for a clean-up. Towns set up committees to inspect 'nuisances' such as dung heaps and open drains, and get them cleared.

There was no idea in the 1830s what caused cholera, and doctors may well have speeded the death of sufferers by bleeding them and thus increasing their dehydration. But 'miasma', akin to the bad air ('malaria') that caused fevers in hotter countries, was perceived as a general cause of disease: breathing damp and dirty air could not be good for anybody. The association of stinking matter and disease led to temporary improvements in the burgeoning towns, and the cholera disappeared. After a few years, it returned: over and over again, as the plague had done in the seventeenth century.

Epidemic disease was frightening and promoted local and national activity, whereas endemic disease had gone with stoicism. Each time the cholera returned, days of prayer were called, but with less and less conviction: and when John Snow[48] established (by statistical studies, first local, then on a bigger scale) that cholera was a water-borne disease, it seemed that attention to drains rather than prayer was required. Snow even had a pub in Soho named after him: beer was safer than drinking water. When appointed Dean of Westminster, William Buckland devoted time to such sanitary measures as became a man of science in an ecclesiastical post. Again, it did not require suspension of belief to appreciate that drinking dirty water could be bad for you. In the 1850s, as water-closets came into more general use, the Thames and other tidal rivers became open sewers in which high tides brought back what had been discharged at low tide. Sittings of Parliament had to be suspended in what was called The Great Stink.

The answer was to build main sewers alongside the Thames, beneath the new Victoria and Albert Embankments, which covered the swampy, smelly areas where mudlarks had sought rubbish to be recycled; and Sir Joseph Bazalgette became one of the first heroes of sanitary engineering. Drinking-water was to be taken in above the tidal limits, filtered and chlorinated before getting into the mains. Victorian energy and common-sense science triumphed eventually over *laissez-faire*[49] in the conquest of this disease, and also of others like the diarrhoea that had carried off so many little children. In 1874 the Registrar General could rejoice in his official report[50] that less than 28 per cent of London children now died before reaching the age of three, though in the same year architects designing schools were reminded[51] that class sizes would get smaller as (not all) children went up the school.

In the same book, the author E.R. Robson advised that the new board schools in England (education became compulsory only in 1870) should have a secular architectural style: he recommended – and used – 'Queen Anne'. His book can be seen as a kind of lay sermon, full of enthusiasm for education with grave suspicion of fanatics from churches who could be useful allies but might be unworthy rivals, misleading the young (and building in churchy gothic).

There was sermonising too in works calling attention to moral and physical health, such as James Kay (later Kay-Shuttleworth) on *The Moral and Physical Condition of the Working Classes* (1832),[52] which were taken up by novelists and those concerned with the condition of England; and there was element of preaching even in Florence Nightingale's delightful *Notes on Nursing*, which brought not only cleanliness but also consideration into nursing – preaching which she reinforced by example as well as precept.[53] These authors were motivated by religious feelings, and Kay dedicated the second edition of his booklet to Thomas Chalmers, the Scottish divine and Bridgewater author who had welcomed the first edition. As the century went on, however, there were more – like Robson on school buildings – whose vision was secular, and who were deeply suspicious of organised religion.

Professional men who thought that way were no longer by the 1870s likely to be charged with blasphemy,[54] as happened to William Lawrence for his medical lectures of 1817. Blasphemy was serious, not just because it offended the ears of Mrs Grundy, but because it offended God; and it was very serious (rather than just offensive) if people really expected His wrath to descend upon those who condoned the sin. Blasphemy put the community or nation at risk of pestilence or disaster.

In 1819 John Abernethy denounced Lawrence[55] for his materialism (he was up-to-date, with new French ideas); and after a tremendous row that deeply divided eminent medical men, Lawrence withdrew the published lectures.[56] But to students they were highly desirable, and pirated editions duly appeared; Lawrence tried to suppress them, but the judge ruled that there could be no copyright in blasphemy, and so Lawrence had no rights in the matter. As a result the lectures were circulated, read and discussed much more widely than if Abernethy had kept out of it; and in due course Queen Victoria appointed Lawrence to be one of her doctors, being more interested in competence than orthodoxy. The dread of blasphemy was lost by the late twentieth century; it was seen as something to do with race relations and mutual respect.

Prayer did not seem to alleviate plague, but scrubbing brushes, drains and water mains did. This common-sense science, transforming expectations in life, represented a threat to traditional religion. People could manage without it. This discovery was just as alarming as anything coming out of evolutionary biology or materialistic chemistry. Huxley said with grim humour that he was prepared to see disease as a punishment from God, but only as a punishment of an ignorant and filthy nation that refused to learn about the inexorable laws of nature and apply those lessons. This God was the player of a game like chess,[57] where we have to infer the rules or die young.

As disease came eventually to be seen as a consequence of germs rather than directly of smells or dirt, so the progressive nature of science was set against the traditional character of religion, the dynamic against the static, in secular rhetoric. Moreover, science was objective, public knowledge, whereas religion came increasingly to be seen as subjective and private. The science of lay sermons was generally common-sense science: the kind that connected disease with dirt and led to effective action against it, or led to developments in technology, as in railways and telegraphs. The arcane and mathematical branches of knowledge, where hypotheses were tested by deduction and quantitative test, were far less accessible – and, as we shall see, they could be seen as built upon faith, in the form of assumptions about the way the world worked.

CHAPTER 9

Knowledge and faith

FARADAY DIED AND WAS BURIED quietly in 1867; a memorial plaque was in the 1990s placed where he would have hated it, in Westminster Abbey – to him a temple of both church and state, God and Mammon. His great contemporary Sir John Herschel was buried in 1871 in the Abbey, near Isaac Newton, whose worthy successor he seemed to be.[1] What they had in common was a deep religious faith, Faraday as a Sandemanian, Herschel as an Anglican. To both of them, pursuing physics – whether electromagnetism or the wave theory of light – was a religious duty: finding out about God through His works.

Francis Bacon was one of Herschel's heroes, and Bacon's phrase 'man the interpreter of nature' resonated for him; in his poem with that title, written in a sonorous Latin metre, the creation was complete only when men of science revealed it:

> Say! When the world was new and fresh from the hand of its Maker,
> Ere the first modelled frame thrilled with the tremors of life,
> Glowed not primeval suns as bright in yon canopied azure,
> Day succeeding to day in the same rhythmical march;
> Roseate morn, and the fervid noon, and the purple of evening –
> Night with her starry robe solemnly sweeping the sky?
> Heaved not the ocean, as now, to the moon's mysterious impulse?
> Lashed by the tempest's scourge, rose not its billows in wrath?
> Sighed not the breeze through balmy groves, or o'er carpeted verdure
> Gorgeous with myriad flowers, lingered and paused in its flight?
> Yet what availed, alas! these glorious forms of Creation,
> Forms of transcendent might – Beauty with Majesty joined,
> None to behold, and none to enjoy, and none to interpret?
> Say! Was the WORK wrought out! Say, was the GLORY complete?
> What could reflect, though dimly and faint, the INEFFABLE PURPOSE
> Which from chaotic powers, Order and Harmony drew?
> What but the reasoning spirit, the thought and the faith and the feeling?
> What but the grateful sense, conscious of love and design?

> Man sprang forth at the final behest. His intelligent worship
> Filled up the void that was left. Nature at length had a soul.[2]

Thought, faith, feeling and 'intelligent worship' were all part of the scientific spirit for that generation, but things were beginning to change.

Belief: a thing of the past?

Reason, tradition and scripture were in classic Anglican thinking seen as the tripod on which belief was built, but we have become apt (since about the 1870s) to think of faith in Benjamin Disraeli's phrase as 'a leap in the dark'. Faith is taken to be irrational, personal and existential, making subjective sense of life, often baffling to others; whereas knowledge seems sober, cool and reasonable, public and accessible, a matter of objective calculation and careful testing. The White Queen in *Alice* – believing six impossible things before breakfast – encapsulated this new attitude: belief starts where certainty ends. Many religious people might appear (to outsiders) to believe 'six impossible things'; in contrast, everybody *knew* that Paris was the capital of France, sodium chloride was soluble in water, and the square on the hypotenuse was equal to the sum of the squares on the other two sides.

Belief seemed to go with religion and fantasy (seen as traditional, feminine and contracting areas of life), while knowledge went with the expanding, masculine world of innovative and universally reliable science:

> Lovers and madmen have such seething brains,
> Such shaping fantasies, that apprehend
> Things that cool reason never comprehends.
> The lunatic, the lover, and the poet,
> Are of imagination all compact.[3]

When we look at narratives of loss of faith we find that it generally resulted from bereavement, betrayal or disaster, or the close critical study of church history or biblical texts, rather than from reading Darwin.[4]

Religious doubt was potent long before *The Origin of Species* (1859) and could account for J.H. Newman's pilgrimage to the Roman Catholic Church.[5] Yet there was an underlying feeling that the manly and questing intellect is not willing to rest in the green pastures of comforting belief, but will forge ahead until it reaches the sunlit uplands of scientific truth: faith ought to be replaced by knowledge, here on Earth through science rather than in Heaven. Although it was good to have faith in one's friends and one's country, there was something unsatisfactory and weak about accepting propositions by faith simply on authority[6] (a question discussed by the prominent politician and editor Cornewall Lewis). Yet, accepting propositions – as set out by creeds, councils or 'articles of faith' – seemed to many to be what Christianity was about. It still does.

When in 1874 John Tyndall gave his presidential address to the British Association for the Advancement of Science (BAAS), meeting in Belfast,[7] he created alarm and despondency among the faithful because he presented a world-picture where gravity and energy, atomic and molecular theory, the wave theory of light and the theory of evolution seemed in principle to account for everything. There was a long way to go before all phenomena could be explained in detail, and no doubt there would be some surprises; but at last men of science had the right end of the stick. They, not the clergy, were now the cosmologists, capable of dealing with the first and last things, the beginning and the end of the world.

From now on, religion would be no more than private spirituality, like Tyndall's own ecstatic experiences on mountain tops. It ought not to be anything more: it could not explain anything. Illness, for example, could be fully explained by germ theory (where Tyndall was a pioneer), but not by religious categories like sin and chastisement. Tyndall and the X-Club seemed, despite their rhetoric of values, to believe in a kind of scientism[8] that accounted for everything in material, causal terms; and the science was securely based on facts, known for certain empirically, and indubitable mathematics.

The 1870s were thus the heyday of what came to be called scientific naturalism, and of Social Darwinism.[9] Auguste Comte's positivism,[10] in which individuals and societies progressed from religious explanations through metaphysical ones to positive science, was an important part of the background – even though it was not popular in detail with scientists like Huxley.[11]

Having faith in science

It was in this atmosphere that in 1879 A.J. Balfour published his first book, *A Defence of Philosophic Doubt*. He had intended to use the word 'scepticism', but that term then was almost a synonym for irreligion. Balfour wanted to apply the sceptical, doubting intellect to science, too readily accepted at face value. He believed that science, like everything else, rested upon faith. That did not mean rejecting it, and he was a friend and admirer of scientists: in 1904 at Cambridge he was president of the BAAS, and in 1920 he was invited by the Council of the Royal Society to be its next president. In the 1920s he became in effect the first Minister for Science in the British government – but to scientism and scientific naturalism he was a redoubtable foe.[12]

Arthur Balfour (1848–1930), now chiefly remembered for the 'Balfour Declaration' that led to the founding of modern Israel, was very well connected. He was educated at home by his mother, Blanche Cecil, sister of Lord Salisbury the deeply religious aristocrat[13] who was prime minister (for the third time) in 1895–1902 and Balfour's patron. Balfour went up to Trinity College, Cambridge, as one of the last 'fellow commoners', a privileged group paying extra fees. He read moral sciences, a new degree subject; his tutor was the utilitarian philosopher Henry Sidgwick, whom he got to know well and who married his sister Nora; he met Darwin's son George and visited the Darwins at Down. He was clearly very able, but working through a syllabus bored him and he got a second-class degree.

With no obvious career, he fell in with his uncle's suggestion that he should stand for parliament, in the safe Cecil family borough of Hertford. His fiancée May Lyttleton died in 1875 and Balfour remained thereafter one of the world's most eligible bachelors. In 1878 he went with his uncle and Disraeli to Berlin, to the conference that divided up Africa; but he was perceived simply as Salisbury's sidekick. Then in the 1880s he became a prominent public figure: the motion that brought down Gladstone after the failure of his Home Rule Bill for Ireland, was drafted in Balfour's house, and in the new Conservative government he was in charge of Irish affairs. In imposing law and order, the man supposed to be a languid aesthete proved astonishingly effective; and in 1891 he became Leader of the House of Commons and spokesman there for his uncle, now prime minister, who sat in the Lords. A lugubrious aristocrat and his witty nephew thus led Britain into the twentieth century, and in 1902 Balfour himself became prime minister.

When Balfour's book came out in 1879, though, nobody took much notice of it. It was well written and amusing, the examples were apt and it required some audacity to be sceptical about science, like Henry Mansel preaching the *via negativa* in religion.[14] But to be taken seriously, perhaps it should have been more solemn. Balfour included little formal history of philosophy, was flippant about German philosophers and urged that there was no certainty to be found, in science or elsewhere:

> Science . . . fails in its premises, in its inferences, and in its conclusions. The first, so far as they are known, are unproved; the second are inconclusive; the third are incoherent. Nor am I acquainted with any kind of defect to which systems of belief are liable, under which the scientific system of belief may not properly be said to suffer.[15]

Science had replaced theology with another dominant dogma but, like everything else, it rested upon beliefs about the world: we are all, willy-nilly, believers.

Balfour's scepticism was not the kind to induce nightmares or existential horrors: he accepted and enjoyed science, religion and things of beauty, and saw inconsistencies as an inevitable feature of finite minds. It seemed to him very wrong that Darwinian evolution, a proper scientific theory, should have been made by devotees into a world-view:[16] he could not accept that natural selection had given us minds. All our beliefs must be corrigible and provisional. Balfour's sceptical method, a version of the theological tradition of the *via negativa*, could not generate religious faith, but it would protect it.

In 1894 Balfour, by now famous, published a second book, *Foundations of Belief*, an introduction to theology. *Foundations of Belief* had also been the subtitle of his first book and the message was not very different. But the parallels between science and religion – one system resting on belief in the uniformity of nature, the other grounded on the existence of God – were made more strikingly; and Balfour attempted to disentangle true science from false naturalism, a monistic and agnostic empiricism that he found incoherent and inadequate. We should not expect our beliefs, which are approximations at best, to be fully coherent; and we should resist any attempts to be dogmatic about the world. Apart from being incoherent, naturalism was gloomy and selfish, as Balfour eloquently put it:

> Man will go down to the pit, and all his thoughts will perish. The uneasy consciousness, which in this obscure corner has for a brief space broken the contented silence of the universe, will be at rest. Matter will know itself no longer. 'Imperishable monuments' and 'immortal deeds', death itself, and love stronger than death, will be as though they had never been. Nor will anything that *is* be better or worse for all that the labour, genius, devotion, and suffering of man have striven through countless generations to effect.[17]

This is indeed the grim world again of Huxley's disciple H.G. Wells, *The Time Machine*[18] and the universe running down.

More things in heaven and earth than scientists dreamt of

Balfour's ideas now attracted much more attention, and in 1896 a group called the Synthetic Society was founded (by Balfour, the Anglican bishops Talbot and Gore, and the Roman Catholic writer Wilfred Ward) to discuss them. Its model was the Metaphysical Society of a generation before, which had brought Huxley, Gladstone and others together in debates, sometimes elaborated in the pages of the review *The Nineteenth Century*, edited by the architect-turned-journalist James Knowles.[19] They had come to feel that they were getting old, that their free and frank exchanges of views were not going to get much further; and they had dissolved themselves. We should never suppose that such groups as this or the X-Club[20] – small groups that last for a relatively short time – are less important than formal and enduring societies. The Synthetic Society in 1909 – not long before its dissolution – privately printed the papers read at its meetings[21] and a list of all its members, prominent among them G.K. Chesterton, Lord Rayleigh, Oliver Lodge, James Martineau and William Temple.

In those papers may be found mirrored Balfour's vision of science, but also new attempts to come to terms with science – for now it was suddenly becoming apparent that it did indeed rest upon metaphysical foundations. Lord Kelvin, one of the founders of 'classical physics' based on conservation of energy, gave in 1900 a famous lecture about clouds over science,[22] noting problems with the æther, postulated as a medium for the waves of light, and with equipartition of energy in gas molecules. In Germany, Max Planck had been wrestling with the way black bodies radiated energy, which did not conform to the predicted pattern; and he invoked the idea that energy came in packets, *quanta*, rather than being continuous.

It had seemed to some, notably Rayleigh, in the generation after Tyndall's that what mattered in physics was the accurate determination of quantities, in a world largely understood in outline; but then in the 1880s and 1890s came unexpected and surprising scientific discoveries[23] – there really were more things in heaven and earth than had been dreamed of. It was all very wonderful and exciting, but the certainties of Tyndall in 1874 now looked more like dogma.

Michelson and Morley, after attending a seminar in Baltimore led by Kelvin, had shown in 1887 that the Earth did not seem to be moving through the æther as one would have expected – a puzzling anomaly, especially since Hertz had that year demonstrated the existence of what came to be called radio waves, and they and light needed some medium to wave in.

Waves without a sea are like Alice's grin without a Cheshire cat. In 1895, Röntgen discovered X-rays; and in 1896, Henri Becquerel showed that uranium was radioactive, stimulating the work of Marie Curie who isolated radium and polonium. In 1897 J.J. Thomson in Cambridge demonstrated in an elegant experiment that cathode rays were (unlike X-rays) composed of minute particles with a negative charge: he called them 'corpuscles', following Boyle and Newton, but they were soon renamed 'electrons' and identified as sub-atomic particles. The reminiscences of Arthur Schuster and of Thomson give us the flavour of these exciting times.[24]

In 1904 Balfour's task as president of the BAAS – an unusual office for a serving prime minister – was the pleasant one of pointing out that there had been a revolution in physics and the science he had learned forty years before was mostly discredited, demonstrating that all knowledge was based upon faith. The classic Cambridge realism of the eminent physicist George Gabriel Stokes – 'A well established theory is not a mere aid to the memory, but it professes to make us acquainted with the real processes of nature in producing observed phenomena'[25] – would not do any more. Balfour did not know of Einstein, quietly undermining classical physics in the Zurich patent office with new ideas about space and time; but one of his sisters had married Rayleigh, J.J. Thomson's predecessor at the Cavendish laboratory, and he was well primed about recent discoveries.

The title of his address[26] was 'Reflections suggested by the new theory of matter' and he began with a run through the history of science, with due emphasis upon Cambridge men. His science was not common sense, however trained and organised: 'the plain message is disbelieved, and the investigating judge does not pause until a confession in harmony with his preconceived ideas has, if possible, been wrung from the reluctant evidence'. Here, as with Davy a century earlier, it is the inquisitor rather than the observer who is the model scientist. The Newtonian idea that atoms were hard and massy had to be revised: electricity, which had been a matter of parlour tricks in 1700, now accounted for matter, which occupied space only as soldiers occupy territory – not shoulder to shoulder, but controlling it with their guns.

We had been preoccupied with the weaker forces of nature; now everything needed to be rethought in terms of complex atoms, 'innumerable systems whose elements are ever in the most rapid motion, yet retain throughout countless ages their equilibrium unshaken'. As in 1800, we see a new dynamic turn given to science in 1900, another repudiation of the 'Newtonian' mechanical world of solid, impenetrable particles.

Balfour disastrously lost the election of 1906 and was out of office until World War I, when he entered the Cabinet first at the Admiralty and then as foreign secretary (and thus responsible for the Balfour Declaration). Just before and soon after the war, he delivered two series of Gifford Lectures at Glasgow, published as *Theism and Humanism* (1914) and *Theism and Thought* (1923), but their ideas will seem familiar to anyone who has read the earlier books. He engaged with the equally aristocratic Bertrand Russell, whose scepticism had led him in a different, anti-religious direction and whose family's politics were opposed to Balfour's conservatism. Balfour believed that his own scepticism (unlike Russell's) was not barren, but constructive.

His writing remained attractive: he told a fellow MP in 1916 that if philosophy was worth anything, there was no reason why it should be dull. On 16 July 1928, when nearly eighty, he addressed a British philosophical society of which he was president,[27] telling them that we lived 'in a world of illusions', urging that 'those who made no effort to get beyond the teaching of common sense should do so in no boastful or self-confident spirit', and adding that 'whenever they found common sense opposed to science, let them throw in their lot every time with science'. Science was incomplete, sometimes untrue, though it was on the way to truth; but 'its ultimate basis, deeper even than experiment and observation, was faith'. Could there be, he asked (amid cheers) a better justification therefore for a philosophical institution?

His nephew, the second Lord Rayleigh, writing an obituary for the Royal Society later published as a little book (1930), reported that Balfour had often wished he had been a scientific man, but had never had enough patience with drudgery and was bored with precision measurements (the first Lord Rayleigh's forte). By the time he died, he was seen as a prophet of the new physics.

Blithe spirits

Balfour's circle not only discovered dogma to be a necessary part of science,[28] thus making room for religious faith as well, but was closely involved with attempts to provide an empirical basis for religious belief, especially in personal immortality.[29] Here the moving spirit was Henry Sidgwick,[30] whose bleak directions for his funeral oration in 1900 were 'Let us commend to the love of God with silent prayer the soul of a sinful man who partly tried to do his duty. It is by his wish that I say these words and no more'. His family were having none of it, and the burial service of the Church of England was used 'without question', although his 'old hope of returning to the Church of his fathers had not been fulfilled'.

Those hopes had depended upon psychical research, the scientific investigation of claims by spiritualists that, through mediums, the dead could be contacted. There was a long history of Swedenborgian beliefs about spirits, and of mesmerism and animal magnetism, in Europe; and in 1850 William Gregory, Professor of Chemistry in Edinburgh, published his translation of Karl von Reichenbach's *Researches*. These included descriptions of auras seen around sensitive subjects, usually women,[31] and spoke of a mysterious 'odyle' (or od) that underlay the phenomena of life.

But the new Spiritualism came from the revivalist districts of the USA to Europe, and especially to Britain. By the 1860s it was verging on respectability,[32] and séances had become a craze; although the mediums, usually young women, were not respectable since the conditions allowed all sorts of groping not usual in Victorian England (or in New England). There were a few male mediums, like the Revd Stainton Moses (of the riddle 'Where was Moses when the light went out?') and D.D. Home, who much impressed Elizabeth Barrett Browning, though not her husband – for whom he was Mr Sludge.

Those who went to séances were usually bereaved (and more people then came into this category, more often); and they might get messages or see extraordinary physical manifestations. The medium might levitate or emit ectoplasm taking the form of the departed. In their popular book *The Unseen Universe*[33] two professors of physics, Balfour Stewart and P.G. Tait, argued for the possibility of human immortality, using the 'principle of continuity': but mediums offered evidence, which might give religion an empirical, even experimental, verification.

William Crookes and a new force of nature

With the wisdom of hindsight it is easy even for the serious historian[34] to describe psychical research as pseudo-science; but the phenomena attracted much attention from men of science. The orthodox Michael Faraday regarded it as pernicious nonsense, and did an experiment using spring-balances to demonstrate that when a group of people sat pressing on a table, it rocked in response to the twitching of their muscles rather than impulses from disembodied spirits.[35]

But, for his admirer William Crookes[36] there were enough facts, witnessed by him and other responsible professional men, to indicate the existence of a new force of nature, psychic force, which scientists should investigate like any other. He submitted his account to the Royal Society, of which he was a Fellow, but after a great row (involving the Unitarian physiologist William Carpenter) it was rejected. In 1871 Crookes published it in his own journal.[37]

He worked particularly with Home (who is portrayed), but also with female mediums. An accordion played music while moving about in a cage, and an arrangement of boards and spring balance indicated strange but powerful forces at work. Crookes realised that the powers displayed in séances were variable and that sometimes nothing happened; but he described series of experiments, under his supervision and that of other men of science, that were inexplicable in ordinary terms. He printed also the correspondence with Stokes, the Royal Society's Secretary and Editor; and his experiments, done before witnesses, were certainly perplexing.

But Crookes was not a gentleman born, though by the end of his long life he rose to a knighthood, presidency of the Royal Society and the Order of Merit. Pillars of the scientific establishment like Stokes were unwilling to take up this kind of research: some mediums were caught out in fraudulent activities. Entertainers claimed to be able to duplicate anything that mediums could do, and remarked how far from streetwise were professional men, whose careers depended on trusting one another.

Meanwhile, Sidgwick had been losing his faith. As an undergraduate and a young Fellow of Trinity College, Cambridge, where he belonged to the exclusive 'Apostles',[38] he had signed assent to the Thirty-nine Articles of faith of the Church of England; but he found himself beset by doubt and by 1869 felt that he must in conscience resign, although there was no legal reason to do so. His upset colleagues in the college kept him on, but necessarily at a lower income and status; and it was repugnance at this event above all that led to religious tests being removed by Act of Parliament.

In 1874 Sidgwick published his classic *Methods of Ethics* where his utilitarian approach severed morality from religion; and at last in 1883 he was the first layman elected to the chair of moral philosophy at Cambridge. He taught political economy, seeking to put it on a more scientific basis; and he and his wife were pioneers of higher education for women. He was very well connected: among his brothers-in-law were Balfour, Rayleigh and E.W. Benson, who became Archbishop of Canterbury. As a disciple of J.S. Mill, he had great respect for the sciences, and all his life could be depended upon to support good causes like reforming college fellowships and widening the curriculum.

The Society for Psychical Research

Hoping for an empirical basis for faith, in 1882 Sidgwick agreed to be the first president of a Society for Psychical Research (SPR), and it was duly founded beneath his aegis. Unlike Crookes, who had heeded advice that his career in pure and applied chemistry was not being forwarded by an interest in psychic force (his motto 'Ubi Crookes, ibi lux' was jokingly supposed to be 'Ubi Crookes, ibi spooks'), Sidgwick made the field seem a proper one for empirical investigation, and collected a distinguished council for the infant SPR.[39]

His own devotion to empiricism and scrupulous verification meant that he never found evidence sufficiently compelling to make him believe that we survive death. The SPR resolved that whenever a medium was caught out cheating, then every séance involving her or him must be disregarded. They also liked evidence to come from 'respectable' witnesses, ladies and gentlemen with nothing to gain by spinning yarns.

Their first major investigation was not concerned with messages from the dead, but with 'phantasms of the living': apparitions of loved ones seen when they were in mortal danger, perhaps thousands of miles away. The three investigators, Edmund Gurney, Frederic Myers and Frank Podmore, published their research in 1886 in two substantial volumes,[40] which also list the distinguished members of the society. The first task was to establish that people really had seen such things; and this required evidence from diaries, letters and remembered conversations, and establishing (by checking longitudes and local times) that the vision was seen at the very moment of crisis. Occasionally more than one person saw the phantasm, and sometimes senses other than sight were involved – they could be touched or heard. Because people might often dream of their beloveds far from home, a census of hallucinations was undertaken to see whether they really did happen more often in disastrous circumstances than otherwise.

The conclusion was that they did, that these appearances were genuine phenomena, and not merely folklore; and that generally there had been no special reason for anxiety at that particular time. The case studies are vignettes giving us a fascinating insight into nineteenth-century life, with friends, husbands, sons (and sometimes daughters) abroad, often in corners of the Empire, facing terrifying dangers at sea or on land. Often but not always they were at the point of death.

But they were still alive: and therefore any explanation must be in terms of telepathy, communication between living minds, rather than of ghosts or spirits coming from the dead. The team therefore at the outset set up experiments with two participants in different rooms, one looking at a playing card and the other trying to write down which it was. Such things had been done as conjuring tricks, but very careful precautions were taken to avoid cheating: and there were some remarkable runs where the results seemed impossible to explain as coincidences.

In more spectacular experiments, drawings made by one participant were reproduced by the other: these remarkable pictures were printed in the book. Ideas and mental pictures seemed to be transferable. The conclusion was that telepathy (experimental and spontaneous) did indeed happen; but it was clearly not straightforward like radio. There were bad days as well as good ones, strong and weak participants, and statistical evidence was the best that could be got. The authors toyed with various hypotheses, but found it hard to come up with anything rigorously testable: though the subsequent discovery of unsuspected invisible rays made telepathy look more plausible in a general way.

Nevertheless, on the whole things got murkier through the 1890s, as the investigators got to work on the spirits. By 1897 Podmore was disillusioned, reporting falsifications by mediums, impositions of 'dramatic unity' by witnesses (memory is a creative faculty, and we all do our best to make sense of our experiences) and the unreliability of data[41] gathered in darkened drawing-rooms and on good days, rather than under strictly reproducible conditions.

The more optimistic Gurney, who suffered badly from neuralgia (face-ache, an indication of the state of dentistry of the day), died of a chloroform overdose in 1888; the circumstances were obscure, but it was almost certainly an accident rather than the suicide some supposed it to be. Sidgwick, after dining (on excellent form) with the Balfours at 10 Downing Street on the night before he entered hospital for an operation for cancer, never fully recovered and died in 1900.

Curious facts, unordered by theory

Myers meanwhile had moved on from phantasms of the living to investigate similar phenomena happening a little later, when the person 'seen' had already died. His book, again in two bulky volumes, was published (appropriately enough) posthumously in 1903, and dedicated to his friends Gurney and Sidgwick who were both also dead.[42] He believed that the evidence for survival was now good enough for his cumulative argument to be convincing, though like Paley's or Darwin's, it was not tightly deductive. The book is interesting not only for its case studies, but also for its discussions of hypnosis, of multiple personality, and of the 'subliminal' unconscious mind.[43]

Myers was a classical scholar, poet and school inspector, and he wrote well if lengthily: he introduced Freud to British readers in the theoretical discussions in his first volume. He noted that under hypnosis people will do and remember things that they later do not recall, or deny; and he saw this as an experimental way of understanding cases of dissociated or multiple personality – much puzzled over in these years, and made into literature in Robert Louis Stevenson's *Dr Jekyll and Mr Hyde* (1886). For Myers, geniuses are on good terms with their subliminal minds, allowing the flow of ideas to and fro. Among more ordinary people, hallucinations sometimes well up from the subliminal mind of the perceiver, especially when in a state of semi-consciousness. But some hallucinations, he was sure, come subliminally by telepathy from other minds; and some from the minds of the bodily dead. Mediums are sensitive to all three kinds.

Myers' strategy was to build up like Darwin a continuous series of cases running through these categories; and it does make striking and curious reading.

But one of the problems was that, in Myers' analysis, each of us is a hierarchy of personalities dissociated to some degree, our conscious selves being a fragment of the whole, mostly subliminal, mind.[44] Which personality survives bodily death is unclear. Darwin had succeeded in making evolution a serious scientific theory because he had a mechanism, natural selection, which his precursors lacked, and not just a series of just-so stories.

Myers could not, even in the epoch of radio waves, X-rays and cathode rays, offer any account of how telepathy, even between the living, might work. The anecdotes of so-far unexplained phenomena were too like stories of miracles for the comfort of hard-boiled scientists – indeed, Myers believed that the SPR could vindicate stories of miracles against Humeans. But curious facts, unordered by theory, cannot be science.

Sidgwick's wife Nora kept up the programme of investigating psychic phenomena, but found little to convince her firmly empirical mind; and although the horrors of the Great War led to many visits by phantasms like the Angel of Mons, and many bereaved seeking contact through mediums, this does not seem to have promoted the work of the SPR. An abridged version of Myers' book was published in 1919,[45] and another version with a foreword by Aldous Huxley (whose grandfather would not have approved) in 1961;[46] and more phenomena hard to explain accumulated. But there was still no satisfactory theory, and without that a science cannot cohere even when experiments are indubitable – and here they were not.

Lodge and others continued to research and write, but the SPR no longer seemed as central to intellectual life as it had. The hope of proving human immortality that way receded, and came to seem implausible. C.D. Broad,[47] Sidgwick's successor in both the chair of moral philosophy and the presidency of the SPR, urged its relevance and importance, but he aroused little interest among clergy[48] or scientists. In the face of the twentieth-century history of wars hot and cold, totalitarian regimes and national survival, the social gospel and its implications began to be more important than the immortality of individuals.

Moreover, even if the truth of some spiritualistic phenomena had been proved, they would not have led to genuine Christianity any more than natural religion had done. But we cannot doubt that in its heyday the SPR met a need in an agnostic age for something beyond common sense, making room for a kind of scientific spirituality – which brought satisfaction to Myers, Crookes, Lodge and others, even if more resolute empiricists, like the Sidgwicks, eventually found it as disappointing as orthodox religion.

'The very hairs of your head are all numbered'

In the 1851 Census in Britain it had emerged that 50 per cent of the population were churchgoers of one kind or another: and this figure seemed shockingly low to most Victorians. The percentage presumably continued to fall, but not disastrously, and population growth helped to keep up the numbers in the pews. The eighty years or so up to 1914 saw in the English-speaking world an astonishing and unprecedented amount of church restoration and church building; and also a surge of interest in missionary activity. Wherever we look in British cities we see new churches that were built, or medieval ones more-or-less thoroughly 'restored', in these years. Right through the long nineteenth century, it would be very difficult to overestimate the importance of public religious belief; and the faith of great numbers of scientists remained more or less orthodox.

This faith showed itself in public concerns, to do with social justice and the condition of the poor, with slavery, prostitution, treatment of indigenous populations in the empire and the Wild West, and with abstinence from alcohol; and also in worship, made livelier by new hymns[49] and slowly increasing formality and ritual, where the Bible was read in various ways but certainly not simply as a straightforward account of a number of facts. The creeds were also recited, but again not simply as a series of things to be assented to, but as an expression of togetherness and a reminder that faith should have consequences, and was not simply a personal construct that made one feel better.

There were many like Kelvin who found no serious difficulty in reconciling their science with their churchmanship, as Faraday and Herschel had done. At the end of the century, the liberal Frederick Temple, sympathetic and understanding about science, became Archbishop of Canterbury;[50] and Charles Gore became Bishop of Oxford and editor of *Lux Mundi*, a volume of essays welcoming evolution and other modern thoughts.[51] For many, if not for Sidgwick, it remained very possible (as it was for Balfour) to continue true to the faith of one's ancestors in a world where intellectual frontiers were shifting.

'Science' for the early Victorians had covered every area of reliable knowledge – the word still does in many other languages – including history and the study of languages; but these were beginning to pose problems at least as acute as those posed by physics or biology. Perhaps for that reason, by about 1900 'science' had been trimmed to its modern meaning; but the question remained how far science should be the model for all knowledge.

Classical physics like Tyndall's had seemed the fundamental science, and Kelvin[52] had taken on Huxley and Darwin by challenging the vast time-scale they proposed. Sciences like biology and geology should, according to physicists at least, be built upon the basic assumptions of physics. Moreover, mathematical physicists ever since Herschel[53] had insisted upon 'a sound and sufficient knowledge of mathematics . . . without which no man can ever make advances in this or any other of the higher departments of science, as can entitle him to form an independent opinion on any subject of discussion within their range'.

Physics was to be popularised *de haut en bas*, and taken on authority. Naturally, to the annoyance of professors of physics, this did not always happen: and one of the most distinguished of their number, Faraday, was ignorant and distrustful of mathematics – but he was generally seen as a chemist, which was how he had been trained by Davy, manipulating things with his fingers rather than symbols in his mind.

Euclidean geometry had been the great example of a deductive system, evolved *a priori*, that fitted the world: the discovery of non-Euclidean geometries alarmed mathematicians, but was not very widely known – the term 'non-Euclidean geometry' in Britain usually meant merely that the theorems were taught or proved in a somewhat different order or manner from Euclid's. The revolution in physics did little to dent the arrogance of physicists; even if they had to admit with Balfour that their science rested upon faith (on assumptions about uniformity, space and time), it was still paradigmatic: the saying attributed to Rutherford, that science is either physics or stamp-collecting, was their general view. My physics teacher was a typical graduate of Edwardian Cambridge when he spoke of 'the less exacting discipline of the more descriptive sciences'. So, while reducing religion or literature to evolutionary psychology might seem wild, reducing chemistry (and then other sciences) to physics seemed highly plausible, especially to physicists.[54]

With Niels Bohr's quantum theory, the arrangement of the chemical elements into the Periodic Table could be explained in terms of electronic orbits; and it seemed that genuine chemical theory must come from physics. Structural chemists fought back against such explanations in principle, and urged that their science was no more reduced to physics than is architecture: buildings and molecules have to be built in accordance with physical laws, but that constraint does not prevent creativity or determine outcomes.

But even to arrogant physicists, the new world of quanta, radioactivity and complex atoms presented alarming puzzles.[55] Received notions not only of space and time, but also of causality no longer seemed to apply, and the indeterminacy associated with sub-atomic particles (even though law-governed) seemed to offer some possibility of reconciling human freedom with the iron rule of physical necessity. But perhaps most significant was the importance of statistical explanation.

James Clerk Maxwell had in 1859 shown how the behaviour of gases, composed of millions of molecules, could be accounted for statistically; and in the same year Darwin's theory of natural selection depended upon informal statistics. For the elderly Herschel, one of Darwin's heroes who had been sent a copy of the *Origin*, it was 'the law of higgledy-piggledy' – mere chance[56] – but for Darwin it was development in accordance with statistical law. It was during the period under discussion, the later nineteenth century, that this feature of the fleeting world – statistics – was captured; and to that we shall turn in our next chapter.

CHAPTER 10

Handling chance

IN LIFE, CHANGE AND CHANCE preoccupy us. Are we victims of fate, nemesis, the iron laws of necessity? How much of our life is fixed by fortune's wheel, mere fluke, coincidence? Is it just good or bad luck? Is our future already fixed, Calvinistically predestined? Or are we free agents, in control of our lives? These are big questions that affect us all.

The superstitious have their charms and amulets, their private rituals when buying lottery tickets; religious-minded people seek some plan, a destiny that shapes their lives and makes sense of their experiences; while scientists wonder, with Einstein, whether God plays dice. Sometimes we interpret an illness or accident impersonally – a result of exposure to bacteria and viruses, or of tumbling downstairs under gravity – and sometimes we see it personally, as something life-changing. Even if we influence nothing else in our lives, how we see that illness or accident will affect how we come through it.

Our insurance company can predict how many of us will fall ill or fall downstairs; their marketing department can predict how many of us will respond to their mailshots. As that great Roman consumer, Horace, said, 'We are just statistics'.[1] Yet, despite our predictability, insurance companies still fail. How scientific is the law of averages?

What are the chances that everything has a purpose?

Huxley's vision of life as a game of chess against an always fair but unforgiving opponent, where we must work out the rules or die, looks not only grim but too purposeful for so chancy a thing as life.[2] We admire those who respond to chance with fortitude, like Job, or boldness, like Nelson. But what were they responding to? Can any spirit be found in a world of chance? Can such an unstable world have anything worshipful about or behind it?

One way to think about chance is in terms of its apparent opposite, purpose. We do not as a rule do things without a reason, though it may not be rational: we may elect to follow our whimsy sometimes, taking a new turning on a country walk perhaps or buying something on impulse. Within some general purpose, our action may not be very carefully aimed: an enemy soldier drew a bow at a venture, and killed the king of Israel.[3] Final causes seemed barren virgins to Francis Bacon, but we look to them to account for our actions and those of other people.

For Plato, for Aristotle and for the great physician Galen it was necessary also to look for final causes in nature and her (or its) works: Nature does nothing in vain, everything is part of a purpose or plan (which occasionally goes awry). To understand at least part of that plan, Aristotle dissected many fish and opened hen's eggs daily to see how the embryo developed; and Galen did experiments, notably on the kidneys and bladder of pigs, revealing how intricately the parts of animals cohered. One can be sure (at any rate, it is a good guiding principle) that all organs have a function, and the job of the physiologist is to go beyond descriptive anatomy to final causes and ask, What's it for?

This reasoning guided Cuvier, one of Charles Darwin's 'two gods' (the other was Linnaeus), in his reconstructions of extinct creatures[4] because every bone must have its place and its function. The notion of the ancient atomists – Democritus, Epicurus and Lucretius – that everything resulted from the random hooking together of atoms seemed utterly absurd to those who studied living creatures. It looked therefore as if chance had to be set against purpose.

Of course, the purpose of a hip bone and the purpose in what we do are two different things, but in a providential view of the world they are just two aspects of God working His purpose out. There was a reason for everything and the realm of science seemed to be to find reasons for things. Science could be applied to things, make them predictable, repeatable, certain – and these were appropriate adjectives for purpose, cause, law, contrivance and design.[5]

Charles Sherrington's classic Gifford Lectures of 1937–8, *Man on His Nature*,[6] were focused on this theme. One-off, improbable events, coincidences, unlikely and local concatenations, these happened all the time in our lives and kept us on our toes: but they seemed out of place in the orderly world of the laboratory, from which everything accidental had been excluded and where events could be repeated at will.

Estimating the odds

There is a military maxim that goes 'Once is happenstance, twice is coincidence, but three times is enemy action' and to sort out chance from purpose involves statistics of this sort. For the eighteenth-century Bishop of Durham, Joseph Butler, probability was the guide to life; as it must be for politicians,[7] those in business and indeed all of us.

We are not always very good at estimating the odds: Beddoes at the beginning of the nineteenth century deplored the way his contemporaries were terrified about mad dogs, though cases of rabies were mercifully very rare, but took for granted a high death rate from tuberculosis and ignored his exhortations to a healthier lifestyle.[8] A thousand to one, a million to one, a billion to one, may seem either highly unlikely or very probable, depending on our state of mind.

Anyway, because the future cannot be foretold with certainty and we can be sure it will surprise us, we clearly have to be guided by forecasts, guesswork, crystal balls or common sense. We might reasonably think that events are unpredictable because the world is so complicated: there are countless people leading their lives, impinging on one another, and countless natural processes going on all around us. Although there are no doubt reasons and causes for everything that happens, surely so many are involved that approximate forecasting is all that can be expected.

Weather is notoriously hard to predict, but it is now much more accurately done than when Darwin's Admiral FitzRoy issued the first public weather forecasts in the mid-nineteenth century[9] – we know more about weather systems, we can see from satellites deep depressions coming across the Atlantic, we have computers. Even so, the weather can still surprise us and all we can do is hope that the day we fix for our garden party will be sunny.

Our ancestors prayed for fine (or at least seasonable) weather and held rogation processions beating the bounds of the parish and calling down God's blessing on the fields. We may still do it, but without the same seriousness because we know enough about weather-fronts not to suppose that every storm or sunny interval is sent especially by God to reward, educate or punish us. Disastrous and unpredictable accidents are still called 'Acts of God' by insurance companies, but we don't really see them that way.

We expect God's word to help us make the best of a washed-out event or a fire-damaged building – helping us to keep cheerful and keep going – but we don't expect Him to change the rules for us. We are much less likely to see providence and design in events than our ancestors, and much readier to see what happens as chance, despite their idea of fortune's wheel.

Not all chance events are the outcome of complex and interacting systems, like the weather. Horse races and football matches have many unpredictable variables, but coins, dice and roulette wheels are simple things on which people also place bets. The early eighteenth-century mathematician Abraham de Moivre was one of the first to deduce the odds for various outcomes of coin or dice tossing, and wheel spinning; and at the turn of the century Pierre Simon Laplace wrote both a learned ('analytical') treatise and a popular ('philosophical') essay on probabilities.[10]

We often remember Laplace as the great determinist: he invoked a Being who knew the position and velocity of every particle of matter in the universe, and thus had the past and the future clear before his eyes:

Yea, the first Morning of Creation wrote
What the Last Dawn of Reckoning shall read.[11]

But clearly that was not the whole story, and in his popular essay Laplace sought to move beyond dice and apply probabilistic thinking to human behaviour, to such complex questions as what majority in juries' verdicts would secure a high rate of conviction for the guilty and a low rate for the innocent. He had to make judgements about how frequently people told the truth, how often they made sensible decisions based upon evidence, and so on. His conclusions, in keeping with his deductive approach, were not based upon any empirical data but on *a priori* assumptions about human nature.

Napoleon made him Minister of the Interior, but soon retired him to a sinecure post in the Senate where he would be less able to apply his mathematical reasoning to social and political questions. We can see why Whewell so disapproved of Laplace, the man who 'had no need of the hypothesis' of God: Whewell came to believe (despite his own eminence in and admiration for applied mathematics) that purely deductive thinkers could not be relied upon for wisdom beyond the realm of pure mathematics.[12] Nevertheless the nineteenth century was the epoch in which laws of chance were worked out, and more empirically than Laplace would have expected.

A collection of facts

It had seemed that law and chance were opposites. Law can be seen as the expression of authority and responsibility, a humane framework for life.[13] The lawgiver was a valuable image of God, who had chosen for the world the best of laws to govern it, like a statesman drawing up a new constitution. Whereas human laws, applying to awkward or perverse persons, are frequently broken, it seemed that natural laws were inviolable – planets or atoms could not but obey them, having no capacity for choice in any case.

In fact, this is not true for what we call scientific laws: no planet 'obeys' Kepler's Laws exactly, because under gravity they interfere with each other; all actual gases are lawbreakers where Boyle's Law is concerned; electronics depends on breaches of Ohm's Law; and living creatures defy the famous Second Law of Thermodynamics (until it catches up with them and they die). Matter seems no better at keeping rules than teenagers are. Laws in science, much lauded since Bacon's day, appear to be idealisations, akin to what Aristotle called 'formal causes' – more like codes of practice than regulations against murder or car-parking.

But to most people before the nineteenth century, just as the reign of law was the opposite of anarchy, so the reign of natural law had no place for chance in scientific explanation. As Whewell put it, "Nature . . . is a collection of facts governed by *laws*"[14] – attributing something to 'chance' simply meant 'Don't know' – whereas Charles Darwin believed that law (a law of nature, that is) came in a category between purpose and chance. Either way, accounting for the weather or for human behaviour was so complex that, though there must be causes for everything, they could not (yet) be calculated; but in simpler spheres like astronomy or inorganic chemistry nothing happened by chance or even unpredictably.

Lies, damn lies and statistics

Darwin's lifetime was the first epoch of 'lies, damn lies, and statistics': the first British census (rather behind many other countries) was in 1801, the British Association for the Advancement of Science (BAAS) in 1833 set up a section devoted to statistics, amid great controversy because of its possible political implications,[15] and in 1839 T.H. Lister, holder of the new post of Registrar-General, published his *First Annual Report*[16] of births, deaths and marriages in England – including data from Wales.

The introduction is fascinating, as a tidy and inquiring mind comes to grips with disorder and variety, noting the problems of collecting data, the variety of religious sects and the danger of erroneous inferences. William Farr contributed to this volume and he became a mainstay of the Registrar-General's office until his retirement in 1879, pondering an ever-increasing quantity of data and drawing conclusions in what were described as vital statistics.

In 1859, the year of *The Origin of Species,* the annual volume for London was three times as large as that for England in 1839, and by then there were weekly and quarterly returns, on sale at $1^1/_2$d and 4d, including data about the weather and the cost of living. By then also, causes of death were being more reliably recorded, and much medical evidence could be found in the returns: thus we find in comments on the tabulated data on 5 September that in the previous week[17] 'Six children died of syphilis, 3 persons died of delirium tremens, 2 of disease, induced by intemperance. These are only a few of the deaths caused by vice'. In the following week is found the even more telling remark that 'About 50 persons are destroyed every twenty-four hours in London by sanitary defects, that we may hope will be to a large extent removed by the Board of Works', which was building a main sewer system. On 28 November, reviewing the week in which the *Origin* was published,[18] the number of births was 1,785, up from 1,548 in the same week a year earlier; but the numerous deaths of small children remind us what a very different world this was.[19] Inductive inferences from statistics, with their political charge, transformed the way people looked at society, and suggested that laws underlay observed regularities.

Just such a 'statistical law' lay behind Darwin's idea that natural selection was the engine for descent with modification, which he called development. He avoided the word 'evolution', because it referred to processes with final causes, like the growth of an embryo. Darwin's law was what Thomas Malthus had called the struggle for existence,[20] by which those (even slightly) better endowed for the niche they occupied would be more likely to survive and leave descendants – who might in turn share this advantageous endowment.

This is not a rigorous law of nature, but a statistical one, telling us what is most likely. Whereas for coins and dice, provided they are not bent or loaded, statistical laws are precise and quantifiable, in other contexts (as here) they are more of a trend, where quantities are variable and uncertain, though they can be refined by more evidence, inductively.

Darwin was hurt that John Herschel (to whom he had sent a copy of the *Origin*) referred to his theory as the law of higgledy-piggledy;[21] and he was unhappy that many scientists, as well as Bishop Wilberforce, regarded the *Origin* as unphilosophical, meaning that it did not measure up to accepted standards of method and explanation, and could be dismissed as hypothetical or speculative. Nobody had ever seen one species change into another, and so Darwin's was not a proper inductive generalisation; and no definite outcome could be deduced from his law and tested by experiment, as the existence of the planet Neptune had recently been predicted from the Newtonian theory of gravity.

Darwin's theory still offends philosophers who believe that scientific laws cannot be accepted unless they can be tested – Karl Popper's criterion of 'falsifiability'. So much the worse for them: testability has to be seen a little more flexibly. Herschel in his scientific activity sought like Newton for a genuine cause, *vera causa*, for every phenomenon.

But just as great poets expand the reader's ideas of language and metre, so great scientists expand our idea of science. Darwin himself made one prediction, that a handsome white orchid from Madagascar would be fertilised by a moth with a tongue ten or eleven inches (over 25 cm) long[22] – which turned out right; and Lyell made another, that if we have a common ancestor with the gorilla, the chimpanzee and the orang-utan, then fossil evidence will be found where they still live, in Africa or Indonesia[23] – which has also turned out right. But undoubtedly where statistical laws are the basis, and history with its particularities is involved in explanations, quantitative Newtonian predictions cannot be expected.

Curiously enough, it was also in the autumn of 1859 that the devout physicist James Clerk Maxwell presented at the BAAS meeting in Aberdeen his dynamical and statistical theory that gases were composed of vast numbers of molecules colliding with each other and with the walls of their container. This idea was not new: Christopher Wren was among those who had tried to make quantitative sense out of it, and in the nineteenth century there were many others. Newton had supposed (like Wren) that the particles must be absolutely hard, made by God so as never to wear or break in pieces:[24] but collisions of such hard particles are hard to imagine. For Maxwell,[25] they were instead perfectly elastic.

Heat is work: the future is cool

Another theory that had made a comeback was Newton's (and his contemporaries') that heat was motion – an idea displaced in the years around 1800 by

Lavoisier's view that heat was a kind of fluid capable of chemical combination – and this prototype of the first law of thermodynamics was seen in the work of Dalton's pupil James Joule, who had heated water by stirring it.[26] In hot water the molecules were moving faster than in cold.

Maxwell developed this idea: but while, in a mass of hot fluid or gas, particles will on average be going faster than in cold, some particles (having just collided) will be going slowly. Where huge numbers are concerned, as in any real state of affairs, we may deal with averages; and Maxwell, Rudolf Clausius and Ludwig Boltzmann developed the theory to account quantitatively for the way gases were found to behave – a wonderful linking of physics and chemistry.

The second law of thermodynamics gave a direction to the whole universe by indicating that heat would always flow from warm to cool bodies, so that eventually everything would be at the same temperature. This would be the end of the world: not a bang, but a whimper. Thus when jars of hot and cold gas are brought together, the whole becomes tepid. Maxwell, like Descartes and Laplace before him, invented a being – usually called his Demon – who was so small that he could distinguish individual molecules, and who was placed between the two jars of now-tepid gas with a frictionless trap door.

The demon let through fast-moving particles from right to left, but stopped slow ones; and conversely allowed slow particles to go from left to right, but stopped quick ones. After a bit, the left jar would be hot, and the right cold; and he would have done this without expending energy, and thus defied the second law. Presumably, being a demon rather than flesh and blood, he could go on and on doing this. What the story indicated was that there was a statistical rather than a strictly deterministic basis for one of the fundamental laws of nature: that physics was not quite as Laplace had imagined it. Darwin was not out on a limb in basing his theory on chance, for chance had its laws after all – though in physics they were stricter.

Divine Providence and insurance

Darwin's thinking had been triggered by reading Malthus, and his theory was also related to the competitive capitalism of Great Britain,[27] the first industrial nation. Those who see science as a social construction, reflecting human preoccupations and relationships, rather than a value-free account of how the inanimate world works, find support in the connection of social and natural sciences through statistics. Here social and intellectual pressures came together.

A hundred years before Darwin was born, John Arbuthnot (satirist and doctor) had in 1710 detected Divine Providence[28] from statistics, in

> the exact balance maintained between the numbers of men and women, for by this means it is provided, that the species may never fail, nor perish, since every male may have its female, and of a

proportionable age. This equality of males and females is not the effect of chance, but Divine Providence, working for a good end, which I thus demonstrate.

He began with coins, where in the long run an approximately equal number of heads and tails would be expected; but then moved from the mathematical to the physical.

An examination of christening records from London (in the famous Bills of Mortality) from 1629 to 1710 showed that in every year more boys were born than girls, which was so extremely improbable on the basis of his calculation for coin-tossing that 'it follows, that it is art, not chance, that governs':

> To judge of the wisdom of the contrivance, we must observe that the external accidents to which males are subject (who must seek their food with danger) make a great havock of them, and that this loss exceeds far that of the other sex, occasioned by diseases incident to it, as experience convinces us. To repair that loss, provident nature, by the disposal of its wise Creator, brings forth more males than females; and that in almost constant proportion.

Arbuthnot concluded that from the resulting equality of the sexes, polygamy is shown to be contrary to the law of nature and justice, and to the efficient propagation of the human race. Steady divergence from the 50:50 chance that might have been expected was evidence, not in this case of enemy action, but of divine dispensation; presumably built into His plan, for God in Arbuthnot's day was more revered for rules than for exceptions. Since the mammalian ovum was not observed for more than a century after this paper,[29] medical men could not provide an explanation of this phenomenon: it remained a curious fact.

It is a feature of the modern world that, in the face of life's vagaries of life, people do not trust in providence: they take out insurance. Buying fire insurance, life insurance and annuities (income for life) was becoming commonplace for Britons in Arbuthnot's lifetime and, in the absence of British statistics, the astronomer Edmond Halley used data from the Silesian city of Breslau (now Wrocław), where the records were good.[30] Insurance companies needed vital statistics, such as Farr later supplied, to make their rates realistic and competitive; and, by the nineteenth century, mathematicians served as actuaries, providing the calculations they needed.

Augustus de Morgan,[31] a Cambridge logician, eked out his salary by this means. Brought up an evangelical, he turned non-denominational with a preference for the Unitarians, and got a professorship at the secular University College, London. These were part-time posts, where the salary depended on student numbers. De Morgan famously twice resigned from his underpaid chair, once when he detected the college being pro-religious in dismissing a teacher, and once when they were anti-religious in making an appointment. It was said of Unitarians[32] that they were deists who had taken shelter, which was unfair to de Morgan and others, and that they were 'loving rather to question than to learn', which he would have seen as a compliment.

Galton's *Hereditary Genius*

If evolution by natural selection were a lawful, statistical process, it ought to be possible to quantify it one way or another. Charles Darwin's cousin Francis Galton was the crucial figure in Britain in measuring humans to provide the inductive base required. Belonging to the much intermarried Darwin and Wedgwood families,[33] he was the spoilt boy in a family of girls and, knowing he was extremely clever, made his name in studying hereditary genius.

Galton travelled in southern Africa and wrote a colourful book of advice for adventurous travellers,[34] first published in 1855. It can be contrasted with the more austere volume, edited by Herschel with a chapter by Darwin, published in 1849 under the auspices of the Admiralty,[35] where scientific inquiry loomed larger. Galton is full of wonderful advice, on clothes and how to keep them dry, through good tips on clandestine surveying, crossing rivers and hunting, to how to manage natives and avoid hostilities. His travels won him a medal from the Royal Geographical Society in 1854 and, like those of Banks, Darwin and Huxley, they launched him into the scientific community – in 1857 he became Honorary Secretary of the Royal Geographical Society, and in 1860 was elected a Fellow of the Royal Society.

In 1869 Galton published *Hereditary Genius: an Inquiry into its Laws and Consequences*,[36] in which he tabulated the families of statesmen, men of science, scholars, writers, oarsmen, wrestlers and others to show how important heredity was. He sought to classify men according to their reputation and their natural gifts, in a kind of meritocracy. His own extended family ramified into the élite professional class, which formed a clerisy or intellectual establishment rivalling the older ruling class based on land and church.

He noted that his men of genius were not effete, but generally big, strong and vigorous; and this reflection leads into his discussion of the 'comparative worth of different races', where he is a right-wing social Darwinian:[37] 'out of two varieties of any race of animal who are equally endowed in other respects, the most intelligent variety is sure to prevail in the battle of life'. In his experience and reading, 'it is seldom that we hear of a white traveller meeting with a black chief whom he feels to be the better man'. There were, he noted, some black men of high ability, but the average was lower: this was after all a statistical investigation. Some Indians, Hyder Ali and Runjeet Singh, featured amongst the great military commanders in his list, with their eminent sons.

But now even the top people, the 'industrious Lowland Scots and northern English' to quote Galton, faced the problems of keeping civilisation progressive, and 'it seems to me most essential to the well-being of future generations, that the average standard of ability of the present time should be raised' – the idea of original sin, our consciousness that we do not do what we ought, is not a sign that we have fallen from perfection, but that we struggle to keep up with the pace of our development from primitive or savage ancestors.

Important matters, then and now, depended upon understanding natural selection: for example, the prudent who postponed having children until their thirties would decline in numbers because their few descendants will be swamped by the offspring of the more prolific; and similarly the Middle Ages were so prolonged, because anyone nicer or brighter than average went into a monastery or nunnery and (usually) left no children to inherit their good characteristics.

Galton ended his book with Darwin's notion of 'pangenesis', a fairly explicit version of the idea then widespread that inheritance was a blending of characters, taken from all the cells in their parents' bodies: he applied his statistical methods to this theory. When in 1892 he wrote a new introduction to the book, he concluded:[38]

> I wish again to emphasize the fact that the improvement of the natural gifts of future generations of the human race is largely, though indirectly, under our control. We may not be able to originate, but we can guide. The processes of evolution are in constant and spontaneous activity, some pushing towards the bad, some towards the good. Our part is to watch for opportunities to intervene by checking the former and giving free play to the latter. We must distinguish clearly between our power in this fundamental respect and that which we also possess of ameliorating education and hygiene. It is earnestly to be hoped that inquiries will be increasingly be directed into historical facts, with the view of estimating the possible effects of reasonable political action in the future, in gradually raising the present miserably low standard of the human race to one in which the Utopias in the dreamland of philanthropists may become practical possibilities.

The answer was eugenics, whereby the unfit in society would be discouraged (or prevented) from propagating their kind, and the fit encouraged to do so. Even a statistical investigation into a complex process could lead to political action.

Speaking of eugenics, we nowadays recall 'historical facts' from Nazi Germany, clearly not the 'reasonable political action' that Galton had in mind and about as far as one could get from a philanthropic utopia; and eugenics seems remote from any spiritual view of human kind. But I was myself a beneficiary of eugenics: in an attempt to get the brainy to breed, university lecturers were paid £50 per child per year, on top of a princely salary (£1,000 a year by 1964): and we got as far as our fourth child before the allowance was abolished.

This was positive eugenics; the negative side, not confined to Nazi Germany, involved measures to stop 'degenerates, throwbacks, idiots, criminals' and other stigmatised people from having babies. Fin-de-siècle Europe (and North America) went in fear of the underclass, was shocked at the ill-health of city dwellers revealed in army medical examinations, and was acutely conscious of race.[39]

Men of science, naturally selected

On 27 February 1874 Galton gave a Royal Institution lecture[40] on 'Men of science: their nature and their nurture', based on questionnaires he had distributed to leading practitioners. Far from being weedy, absent-minded, bespectacled geniuses, his sample of eminent scientists were energetic, healthy, vigorous, independent and persevering. These strong points were coupled with an interest in mechanisms – evident from childhood – and an innate taste for science, an abiding passion sometimes expressed as love of nature and desire to learn her secret. Men of science were unemotional, interested in facts and theories more than in persons, and manly to the point of being anti-feminine: their fathers had influenced them much more than their mothers. At school, they had valued any free time (which does not fit with our cramming theory of education).

Galton concluded his rather schoolmasterly oration with this vision of a church scientific:

> The majority will address themselves to topics nearly connected with human interests, a few only will turn to science. This tendency to abandon the colder attractions of science for those of political and social life, must always be powerfully reinforced by the very general inclination of women to exert their influence in the latter direction. Again, those who select some branch of science as a profession, must do so in spite of the fact that it is much less remunerative than any other pursuit . . .[The] gigantic monopoly [of Church and classics] is yielding, but obstinately and slowly, and it is unlikely that the friends of science will be able, for many years to come, to relax their efforts for educational reform . . . it is to be hoped that, in addition to the many new openings in industrial pursuits, the gradual but sure development of sanitary administration and statistical inquiry may in time afford the needed profession. These may, as I sincerely hope they will even in our days, give rise to the establishment of a sort of scientific priesthood throughout the kingdom, whose high duties have reference to the health and well-being of the nation in its broadest sense, and whose emoluments would be made commensurate with the importance and variety of their functions.

Galton's prose (as written, and no doubt as spoken) lacked the punch and passion of Huxley's, but the message was much the same: science was the new spirituality and our best hope for a healthy and progressive society. Galton distributed offprints of his lecture, repaginated, postage ½d, for those who had not heard him.[41]

We tend to accept that social and scientific Darwinism cannot be disentangled, and thus evolutionary theory can be seen as socially constructed. Whatever its origins, its powerful hold on our thinking seems to indicate that it must reflect what goes on in nature, like the 'hard sciences' of the inorganic world. Certainly, Darwinism is seen by most people as established beyond reasonable doubt, so it was not just historians who were startled in 1981 when Donald MacKenzie published his *Statistics in Britain, 1865–1930: the Social Construction of Scientific Knowledge*.[42] This indicated how based in eugenics (socially invented and not just socially driven) were the statistics of Galton, his disciple Karl Pearson and Ronald Fisher, who used them to show the compatibility of Darwinian natural selection and Mendelian genetics.

The bell curve – used by Gauss to correct astronomical observations – had been adopted by Quetelet for studying human variation. Galton took this much further, making important contributions to statistical methods now routine in biological, medical and social sciences. For the beady eye of the eugenicist, it is not the peak but the rim ends of the bell which are interesting – they show the highly superior and inferior. Galton and Pearson were both driven men, who had lost their religious faith and found a substitute in science. To us, their utopian eugenics seems hopelessly class-based, racist and harsh – 'How shalt thou hope for mercy, rendering none?' as the shocked Duke asked Shylock.[43]

But down to 1939 eugenics retained its appeal, not only to agnostics in search of a new and naturalistic faith, but also to churchmen like Fisher and many others, as revealed in Peter Bowler's *Reconciling Science and Religion*[44] where 'modernist' clergy welcomed evolutionary theory (in a more directed and less chancy form) and eugenic social policies. Law and purpose were reconciled for Archbishop Charles D'Arcy of Armagh, Bishop E.W. Barnes FRS of Birmingham, Dean W.R. Inge and his successor W.R. Matthews of St Paul's, and others, in the belief that God's creation was a continuous process, working through the struggle for existence, which seemed ruthless but was really to be understood as weeding. Christians had a duty to improve the world and artificial selection of humans (as in stockbreeding) was an important part of this – though Nazi-style compulsion, via sterilisation and 'euthanasia', was going too far.

Nature, red in tooth and claw – but God?

This brings us back to where we started: even if we can discern law rather than higgledy-piggledy chance in the process of evolution, does it reveal the purpose of a loving God the Father behind the scenes? Tennyson, writing before Darwin, clearly posed the problem:[45]

> Man, who trusted God was love indeed
> And love Creation's final law –
> Tho' Nature, red in tooth and claw
> With ravine, shriek'd against his creed –

Darwin himself corresponded with his American admirer Asa Gray of Harvard about this problem. Gray, a botanist who had found Darwin's theory very valuable in accounting for the distribution of plants, did not find that it shattered his religious faith, being tested and strengthened in the American Civil War with all its horror.

He wrote a review of the *Origin*, subsequently published in the USA and then by Darwin in Britain, urging its compatibility with the Christian faith,[46] setting off a debate by correspondence in which Lyell and Wallace also joined – Lyell being much impressed by and endorsing Gray's view that

> The whole course of nature may be the material embodiment of a preconcerted arrangement; and if the succession of events be explained by transmutation, the perpetual adaptation of the organic world to new conditions leaves the argument in favour of design, and therefore of a designer, as valid as ever; for to do any work by an instrument must require, and therefore presuppose, the exertion rather of more than of less power, than to do it directly.

For Darwin, the course of nature did not seem predetermined and he could see no evidence of intelligent design: only evolutionary bushes rather than a tree with humans at the top; and a bloody, chancy process whereby animals such as barnacles and tapeworms went down in the world, crocodiles and dragonflies stayed much the same, and primates went up – though the very language of up and down was unacceptable to him. Extinction was the long-term future.

In an optimistic and progressive era, few of Darwin's contemporaries shared this comfortless vision of law-governed process without any overall direction: rather more people, faced with brutal humans and all the animals and plants that captured prey and caused pain, found it as hard as he did to believe in a benevolent and all-powerful God. Darwin could delight in observing nature, writing about creatures almost (as his family said) in the language of advertisements – exquisite structures, beautifully adapted[47] – but most people craved an overall plan rather than the naturalistic and lawful, but contingent, scheme that sufficed for him.

God finds a less nebulous way

Laplace, whom we met as a determinist, atheist and statistician, was also famous for publishing the nebular hypothesis,[48] according to which the Sun and planets had solidified from a whirling mass of gas. This version of creation by natural law was taken up in *Vestiges* (1844),[49] and became rather more respectably known in 1850 in J.P. Nichol's effusive *Architecture of the Heavens*,[50] with its allegorical plates by the romantic Scottish artist David Scott and its wonderful pictures of what could be seen through Lord Rosse's giant telescope, a 72-inch reflector built on his estate in Ireland and a wonder of the age. The spiral nebulae thus revealed seemed at first to confirm the hypothesis of solidifying gases; but the resolution of other nebulae into star clusters weakened it.

Nevertheless, and especially as Ron Numbers has shown in the USA, the nebular hypothesis survived through the nineteenth century, and played a major part in weaning people away from biblical literalism where the creation of the Earth, Sun, Moon and planets was concerned. In 1865 William Huggins, later President of the Royal Society,[51] observed from their spectra that some nebulae were indeed composed of hot gas rather than stars; and this seemed an empirical confirmation of the hypothesis, at a time when nobody knew how far away most nebulae were – they became 'island universes' only in the twentieth century.

The success of the nebular hypothesis meant that anyone acquainted with science was aware that there was a theory, which accorded with observations and had survived some tests and criticism, to account for the solar system and the starry heavens beyond it. And few found much difficulty in accepting that God might have worked this way in creating them.

This process might seem chancy – all sorts of figures had to be just right to bring a world out of chaos – but it was less gory than natural selection. We clearly lived in a world of laws, long established: if we tried to break the law of gravity, we would get hurt – God would not send angels to buoy us up. His power is seen in unvarying action rather than in special providences and miracles.

Nevertheless, inorganic and organic evolution – for most people, involving both a plan and an end, and thus reconciling chance and purpose – had become a feature of their world-view by the early twentieth century.[52] So had statistical thinking, more or less sophisticated. Government depended upon statistics, opinion polls were becoming a feature of life, intelligence tests were placing people on bell curves; and the problem of understanding events against a background of statistical generalisation was with us.

We are back where we started, with subjective experience being set against the objective laws of science – two ways that must be painfully if profitably combined. And to help with such problems, since the mid-nineteenth century there has been growing a 'church scientific', whose prophets included not only Huxley and Galton, but also the devout Faraday and Lord Kelvin. The growth of this learned clerisy, scientific sages keen to guide the unlearned into the paths they should follow, forms the next part of our strange, eventful history.

CHAPTER 11

Clergy and clerisy

FOR GALTON AND SOME OF his contemporaries, scientific naturalism was a kind of religion, encouraging fervent endeavour and a kind of spirituality. They aspired to a scientific utopia, in which pain, disease and poverty would be first diminished and ultimately abolished. Many in our day share the same aspirations. To Thomas Huxley, science had never done anybody any harm, whereas religions were characterised by witch-hunts, inquisitions, intolerance – which might lead to war – and ignorance, which went with neglect of medicine and sanitation.

Saints such as Cuthbert felt the need to escape to a small island or hermitage; and Tyndall's pantheistic experiences came to him alone on mountain tops. Some religion may be what men (and perhaps women) do with their solitariness: but that is a very narrow definition, leaving out most activities of most religions – which are social. So it is with science. It is public knowledge. Do the parallels extend any further? When Galton spoke of 'a sort of scientific priesthood' it was more than a figure of speech. Does science have its clergy? Its saints, even?

A profession of faith

Some Christians, the Sandemanians and the Quakers for example, have no clergy (except in Madagascar, where I have met a Quaker minister); but generally in religious bodies there are duties – liturgical, pastoral and administrative – that are seen to require a ministry set apart. In the Christian tradition this means some kind of ordination: the vocation of a would-be priest or minister is tested by the Church, and after due preparation the candidate publicly makes vows and is blessed by other clergy. In churches that have bishops, they alone can perform ordinations, in a succession going back to the apostles. By the year 2000, in all but the most conservative congregations (and churches are unusually conservative), women were accepted for ordination too. After all this, there was meant to be no doubt about the person's clerical status; and normally only gross immorality could lead to unfrocking.

Despite this, many in the medieval world could claim 'benefit of clergy' because they were in minor orders with few obligations, or simply because they could read; eighteenth-century France had its *abbés* without serious duties or vows; and now there are mysterious *episcopi vagantes*, wandering bishops, who ordain people – and even websites do the same. But for many centuries it has been reasonably clear who are the clergy and what is expected from them. They have been paid (the parson's parish in England can be described as a 'living') and, along with lawyers and doctors, they were the first 'professionals', trained in the medieval universities.

The clergy took responsibility for education, for universities were religious institutions; and the 'public schools' of England were normally headed by clergymen, whose whole career might be spent in teaching. The words 'profession' and 'professional' have undergone a great change: once they meant someone who professed something. We remember John Wesley's distinction between true believers and mere professors. Clergy had at ordination or installation publicly professed their beliefs and their readiness to serve. In return for social position and a stipend (often exiguous, sometimes very comfortable), they were expected to behave better than ordinary mortals, to set an example.

Recognised training

The same went for medicine and law, and for the military, where commissioned officers were expected to behave as gentlemen. Here again – although there were quack medics, attorneys of dubious competence and mercenary soldiers – yet the genuine doctor, lawyer or officer was publicly recognised by his peers and had accepted professional standards. The professions were allowed a measure of self-regulation, with their own courts or tribunals to maintain standards.

In the early twenty-first century, stroppy clergy aspire to professional status defined in the way typical of the age, claiming a salary comparable to a head teacher's – though others might wonder just what their special training and competence is, compared to that of a doctor or lawyer, and see clergy as quintessential amateurs (and, often, social climbers). In contrast, we see accountants, teachers, social workers and even academics as expert professionals. Some of these activities are rather less self-regulating, but the older professions have been brought increasingly under public scrutiny; and, in particular, illegal activities (as we see in cases of paedophilia) can no longer be treated as infringements of a professional code, dealt with by arcane disciplinary procedures. New and old professions involve formal training and elements of public recognition, in differing degrees.

In contrast to the professional, the artisan or skilled worker learned his craft, trade or mystery by apprenticeship: serving his time with a master craftsman, often his father or uncle, but by the nineteenth century perhaps in a great workshop or factory where the family would pay a large premium to get him accepted. This was how medical practitioners, apothecaries and surgeons, were trained in the eighteenth century and beyond; and it was the way the engineers of the industrial revolution learned their trade. Later generations aspired to professional status, though, in Britain and elsewhere.[1]

Encouraging natural philosophers

Where did the natural philosopher, the man of science, fit into this framework of professions, crafts and trades? Davy broke off a medical apprenticeship; Faraday served his time and became a bookbinder, before embarking on another, informal apprenticeship with Davy. Both thus moved decisively from artisan to gentleman status: science was often a vehicle for social mobility, as the Church had been, but there was no point at which Davy nor Faraday publicly qualified as men of science. The wealthy Banks had gone up to Oxford just like well-off or aspiring young men who went into public life, the Church, medicine or the law: he found a tutor to teach him botany, but learned much of his science on his voyages.[2] Wollaston and Young were university-trained doctors, Buckland was a clergyman.

In science, there was no pattern of education – nothing like ordination or being called to the bar, or receiving a commission in the forces – and there was no professional body to set standards. There were also few ways of earning money. Members of the scientific community shared a strong intellectual and practical interest in the world, but they trained and supported themselves in all sorts of different ways.

George Johnston, doctor (trained by apprenticeship and then at Edinburgh) and eminent natural historian of Berwick upon Tweed, emphasised the difference between his life as a professional man and that of gentlemen, whose income came from land or investments and whose time was their own: medicine not only left him little leisure, but brought distracting grief and cares.[3] Johnston was in 1843–4 a founder of the Ray Society, publishing books on natural history for members at an affordable price; and he was a prominent member of natural history societies in north-east England and the Scottish borders. Joining a society was the readiest way to become a man of science. (A woman of science, as in respectable churches, had to remain in the congregation until the late nineteenth century.)

A group who chose to be nameless and met in London in the 1780s in the Chapter Coffee-House, and later in the Baptist's Head Coffee-House, kept unusually full minutes of their meetings. These have recently been published[4] and give a wonderful impression of the discussion that went on, sometimes very inconclusively. Would-be members were proposed and might be blackballed; approved visitors could be brought along. This was the time of Lavoisier's reform of chemistry, which was avidly talked about and mainly rejected; and there were also more practical matters, to do with industry and medicine. The short history and the informality of the institution do not mean it was unimportant; many of its members became Fellows of the Royal Society, or distinguished themselves in other ways.

Learning had once been the province of the Church of England, but the days when the parson might be the only educated person in his parish were coming to an end. By this time, with the success of dissenting academies, there were well-read dissenters prominent in scientific activity – Priestley and his fellow-Unitarians were notable but Quakers were also prominent[5] – but also in the Lunar Society of Birmingham,[6] which was composed of industrialists such as Matthew Boulton and Josiah Wedgwood, the polymath doctor Erasmus Darwin, Priestley himself, and others.

The Lunar Society met less formally than the coffee-house society, being essentially a gathering of friends to dine and talk (business and science) when the moon was full to light them home. Many of these also became fellows of the Royal Society,[7] but even that body was hardly a professional group. It was more like a club, among whose members were a small minority of active researchers who had actually published a scientific book or paper.

In France, Prussia and Russia, academies of sciences provided a focus for aspiring men of science. Election as an academician meant an income, a position, and membership of a body important in government – giving some of the status of a senior civil servant. But positions were few, and the aspirant had to wait for some dead man's shoes to step into. Members of academies exercised some patronage and had a role in the educational world, especially in France after 1789,[8] where science was coming to play an important part in the university, the *hautes écoles*, the Museum of Natural History[9] and the Conservatoire des Arts et Métiers. Academicians also advised nationalised and private industry.[10]

Moreover, the Academy could and did express a collective view, like a church synod, clarifying scientific doctrine and orthodoxy, and warning the public about perpetual motion machines, animal magnetism and other scientific heresies – later, in France, to include evolution and atomism. In the turbulent 1790s, with the suppression of the established Roman Catholic Church, men of science in France really had begun to function as a profession, and even perhaps as a church scientific. The Academy was a collection of sages, some of whom at least would be as ready as bishops with opinions on important matters.

The Royal Society

In Britain it was different. There were university posts in medicine, available to chemists as well as anatomists and physiologists, and much science was done by medical men, in their practices or in academic posts. But the Royal Society had consistently refused to give its collective approval or disapproval to anything;[11] and, although papers in its *Philosophical Transactions* were refereed, the journal carried a disclaimer to the effect that publication did not mean endorsement. Fellows were admitted, as in other clubs, by election after being formally proposed and seconded; and, in Banks' reign, it was the convention to consult the president at the outset, so that inappropriate persons were not put forward. Fellows who knew the candidate personally, or knew his work, would also sign the proposal form; and when it had been displayed for a time, it was voted on. If elected, rather than receiving a salary like an academician, the fellows had to pay a subscription.

As the annual membership lists show, it was a good club, full of professional and landed gentlemen, all with an interest in natural knowledge. Admirals, dukes, judges and bishops could meet prominent medical men, polished intellectuals from Oxford and Cambridge, and men from humble backgrounds like Watt, Davy and Faraday. Banks made sure, as the founders of the society had had to do in the 1660s, that it could not be perceived as subversive. He maintained excellent contacts with the government and became a privy councillor,[12] but was careful to avoid party politics. His successor, Davy, was one of the founders of the Athenaeum Club (making Faraday serve as secretary) for intellectual members of the establishment, as part of a strategy for making the Royal Society more like an academy of sciences.

Only when Davy became president in 1820 was there a majority on the society's Council who had actually published a scientific paper or book. Nevertheless, he believed that the best person to succeed him would be Sir Robert Peel, who instead went on to become Prime Minister. The reform of the Royal Society into a body solely of eminent men of science happened, curiously enough, under the presidency of the Duke of Sussex, one of Queen Victoria's (formerly wicked) uncles, who defeated the brilliant physicist John Herschel[13] in a closely contested election in 1831. It was by no means clear that the best person to preside over the society should himself be closely involved in scientific research.

Charles Darwin set out on HMS *Beagle* as an ordinand, expecting – like Gilbert White or his own friends Leonard Jenyns and William Darwin Fox – to be a country vicar immersed in natural history; he returned determined to devote himself full-time to science. He could do so because his father was wealthy and he did not have to worry about finding a job. If he had not been so fortunate, there were few posts available. Davy and then Faraday had as performance artists in science[14] supported themselves and the Royal Institution; the Astronomer Royal received a salary, and government 'pensions' were available under the patronage system for impoverished intellectuals, including Dalton and Faraday[15] – who in the 1830s abandoned 'professional' science (analyses undertaken for a fee) in order to work on electromagnetism.

Instrument-making, pharmacy, surveying, photography and electricity used scientific knowledge, but there were not many full-time positions at managerial level in British industry for men of science. As the nineteenth century went on, however, factory inspectors, analysts to check water and food for pollution or adulteration, consulting engineers and chemists were required – Crookes supported himself that way. Otherwise, science had to be done in the leisure afforded by work in some other profession: and often men of science were ordained.

The Church of England worked through patronage, and no doubt a congenial living would have been found for a gentleman so agreeable and well-connected as Darwin, with his Cambridge degree. Dissenters had to find a congregation that would employ them, and might well find life more strenuous. Because of the Church's connections with the educational system, it was possible for Buckland to move from his professorship of geology to the Deanery at Westminster. Of Darwin's Cambridge patrons, John Stevens Henslow was Professor of Botany and vicar of a neighbouring parish; and Adam Sedgwick became a prebendary of Lincoln Cathedral as well as Professor of Geology. The astronomer Temple Chevalier was similarly both Professor of Mathematics in the infant University of Durham and vicar of a nearby parish.

What can you teach a clergyman?

Sedgwick in December 1832 gave a famous address on liberal education, in the chapel at Trinity College, Cambridge. It was duly published[16] and later editions had, as an appendix, a prolonged denunciation of *Vestiges*. In response to criticism of the education offered in the university, he urged the importance of character building, of shared rather than specialised knowledge, and of natural theology:

> In speaking of the laws of nature and of the harmonious changes resulting from their action, in spite of ourselves we fall into language in which we describe the operations of intelligence . . . What are the laws of nature but manifestations of his wisdom? . . . Indications of his wisdom and his power co-exist with every portion of the universe.

In Cambridge,[17] the focus for honours was on mathematics, in Oxford on classics and philosophy; but many undergraduates like Darwin read for a broad ordinary degree. They could find time to walk in the gardens with Henslow on summer afternoons, go on his field-trips into the fens, and accompany Sedgwick on his geologising in the Welsh mountains – or they might choose to row, or play cricket or football, because organised sport was in the 1820s becoming a feature of university life, as the memoirs of Charles Wordsworth indicate.[18]

That shared liberal education, its moral values and classical tags, gave gentlemen a common culture, with room for both spirituality and information. The idea was that any specialised study of law, medicine, divinity or science would follow, being built upon a foundation of sound traditional learning and churchmanship. Only from the 1850s did the German pattern (of specialisation within an institution dedicated to research and innovation) begin to prevail. Frontier science with its uncertainties, constant upsets and changes seemed very unsuitable for young minds; established science was all right, and mathematics at Cambridge included papers labelled 'Newton'. Most science came in a context of natural theology and the suggestion that clergymen would show bias or a closed mind in their teaching would have been rather shocking – Darwin, intending to be ordained, would have been a very typical student.

Darwin, on leaving Cambridge in 1831, would have been expected to do a bit of reading before being ordained in the summer of 1832, but that was all. There was scant professional education available for the clergy in his day. The new University of Durham found its niche after 1832 in providing theological training for ordinands at a time when Oxford and Cambridge did not offer undergraduate degrees in theology, but only BD and DD degrees for published work by scholarly clergymen in mid-career. Bishops expected only that ordinands at Oxford or Cambridge would attend the professors' lectures.

Anglican clergymen were well-educated gentlemen, but not specialists; nor did they have a lengthy training – as in the Roman Catholic Church then, and quite generally now – that separated them from their lay contemporaries. Much the same was true of other clergy, taught in dissenting academies and then later at University College, London, along with laymen.

The preparation of Anglican ordinands began to change in the second half of the nineteenth century, and one of the moving spirits was Samuel Wilberforce, Bishop of Oxford. A son of the eminent evangelical William Wilberforce, of anti-slavery fame, he was snobbish and self-assured, and by no means popular with all his clergy. He acquired the soubriquet 'Soapy Sam' for his unctuous way with the Oxford liberals who published *Essays and Reviews* in 1860. Yet he was a thoroughly energetic moderniser of the church, a high-churchman (though no Puseyite) who expected much more of ministers than most people did.

The bishop's palace was at Cuddesdon, near Oxford, and there he set up a theological college, so the clergy would really know their job. As by-products of this quasi-monastic seclusion, each cohort of emerging parsons had absorbed the place's ethos and they all knew each other very well. Wilberforce kept his clergy up to scratch. He took advantage of the railways, visiting parishes and asking incumbents about their activities. He was not unique in this: it was in the spirit of the times and it transformed the clergy into something more like officers in an army. They became more distinct from other people, shepherds set apart from their flock. In a world of rapidly increasing specialisation, the clergy were becoming more of a profession, in the new sense of the word: those with a common training and expertise, undertaking specific duties.

Coleridge and the clerisy

In the 1820s Coleridge, the poet and philosopher, had pondered the relation of church and state.[19] How could you have an established church in a country that had given full civil rights to Protestant Dissenters and Roman Catholics? The Reform Act of 1832, which gave votes to the comfortably off in the industrial cities, made the question acute because Parliament legislated for the Church of England. What if the Commons contained members from denominations hostile to the Church? It already did, and the situation provoked a strong reaction from Oxford men – Keble, Pusey, Newman[20] and Hurrell Froude – who wanted the Church of England to look to its medieval roots, when it really was the national church. They detested dissent and the Anglican evangelicals sympathetic to it.

Coleridge in contrast hoped for a broad and inclusive church, avoiding what he called bibliolatry (worship of the literal text of the Bible). In medieval times, the clergy were the educated class – a clerisy – with a duty to pass on their learning and provide leadership; now, Coleridge urged, those same duties belonged to the educated, who should see themselves as a clerisy. Ordained ministers would be only a part of this clerisy, set apart only for their particular duties, in a new, liberal and comprehensive vision of a national church. Coleridge's views resonated down the century.

Then in June 1833 he helped to define another segment of the clerisy, and make it aware of itself. At the third meeting of the British Association for the Advancement of Science (BAAS), in Cambridge, he took them to task for calling themselves 'philosophers', the shorthand of the day for 'natural philosophers'. He believed that a philosopher was engaged with metaphysics, logic and ethics: for members of the BAAS a new word was required, though he did not offer one. It was Whewell, another great wordsmith, who proposed 'scientist' by analogy with 'artist' and subsequently used the word in a review.[21]

It did not catch on for half a century or more: Faraday, for example, regarded himself as a natural philosopher, thinking broadly about how the world worked; and most others preferred the term 'man of science' – which lasted into the twentieth century, until there were enough women to make the phrase inappropriate. But from 1833 the word 'scientist' existed and there were more and more people who conformed to our idea of what a scientist is.

The British Association

The British Association was the result of pressures at home and a foreign example to follow.[22] Davy, Wollaston and Young had recently died, strengthening the cantankerous Charles Babbage's notion that science in England was in decline in 1830[23] – scientists have often pointed to such decline and urged the need for more funding and honours. Those active in Scotland, Ireland and Wales, and in the English provinces in Bristol, Manchester, Liverpool and Newcastle, felt that the Royal Society with its London base did nothing for them: Dalton, for example, long resisted pressure to join.

In Germany, still a congeries of relatively small, independent states, men of science – if they were to form a critical mass – needed to keep contact across borders. Under the leadership of Lorenz Oken (a natural historian so brimming with the *Naturphilosophie* of Schelling and Hegel that the Ray Society's translation of his book was highly controversial)[24] they held public meetings each year in a different state. The various bureaucracies and police forces were at first alarmed about this, but they soon came to accept that the *Naturforscher*, or men of science, were harmless and might be useful. Indeed, states came to compete for able science professors, keeping up with each other's investment in universities as in opera houses.

Foreigners were invited to these gatherings: David Brewster of Edinburgh, inventor of the kaleidoscope and admiring biographer of Newton, and J.F.W. Johnston, a Durham chemist trained in Glasgow, were among those who went and were much impressed by what they saw. The meetings attracted large numbers and were open to the public. As a result, there was great enthusiasm for science amongst visitors and government officials, and it was directed into appropriate channels. We could even say that these occasions provided a 'form of the Mass in which the audience were themselves identified and uplifted as members of a general community'.[25] The meetings were an important catalyst in transforming Germany into the world's leading scientific nation by the end of the nineteenth century.

Brewster and Johnston were both on the fringe[26] of the British scientific establishment. For a provincial association to succeed, it needed the participation of major figures from the Royal Society and the English universities. Vernon Harcourt, son of the Archbishop of York and himself a Yorkshire vicar, called a meeting at York in 1831; and enough people came to make a go of it. Buckland's discoveries in the Kirkdale cavern and the abundant fossils from the shore at Whitby had aroused local interest, and the Yorkshire Philosophical Society's museum in York had a good collection of fossils. John Phillips,[27] keeper of the museum, wrote a two-volume work on the geology of Yorkshire[28] which had a most distinguished list of subscribers, headed by the archbishop. The first volume was dedicated to his uncle William Smith, who had had the idea of identifying strata by fossils; the second to Harcourt and Sedgwick. Lord Fitzwilliam presided at the meeting, Harcourt being Vice-President, and Phillips one of the secretaries. The meeting propagated a liberal rather than a literal interpretation of Genesis, against 'scriptural geologists' like Dean Cockburn (also of York) and the Revd George Young,[29] who had also published on Yorkshire geology.

There were plenary sessions, and 'sections' for the various sciences; and a good time seems to have been had by all. Buckland, one of whose family had just died, could not go: but he sent best wishes, and invited the gathering – now named the British Association for the Advancement of Science (BAAS) – to come to Oxford the next year, when he duly presided. By the end of the 1830s, meetings had taken place in Cambridge, Edinburgh, Dublin, Bristol, Liverpool, Newcastle, Birmingham and Glasgow, and the association was becoming one of the great institutions of Victorian Britain.

Cities vied with one another, making bids for the association and tempting it with libraries and museums to show off their intellectual life. Eminent 'gentlemen of science' joined Buckland in supporting it: they could be found on the platform and would hold forth, not only to the sessions, but also to (sometimes huge) congregations of working men in the evenings. The shy Herschel as early as 1837 disliked the mutual backscratching that went on:[30] 'I should most gladly have some reason for avoiding the abominable speechifying & flummery of the September meeting of the British Association', calling it also 'that *treacly* affair'; but it soon became the thing to do, to attend each summer, to meet up (perhaps in informal dining groups like the Red Lions) and be seen there.

Public relations were crucial to the meetings: the summer was the 'silly season' when law courts and parliament were not sitting, fashionable people had left London for the country or the continent, and newspapers (expanding with steam presses, railways and increasing literacy) needed news. The president's address usually pointed to the achievements of British scientists and engineers (many there beside him) and called for more support for science. It would be very widely reported, with perhaps the full text; and so would the lectures of the prominent men of science. There would be receptions and dinners, and local worthies, including scientific men, would get the chance to meet those with major reputations.

Local men would read their papers and might attract the attention of the great, as did James Joule of Manchester, whose presentation on heat caught the ear of William Thomson, later Lord Kelvin and the doyen of classical physicists. This led to collaboration, to honorary degrees and to Fellowship of the Royal Society. It was thus possible to rise from the congregation and join what Herschel called the eminent 'mutual be-butterers' on the platform. Women came to the meetings, first as wives and daughters, soon in their own right; but it was rare for women to present papers, though they might well have been involved in the researches announced by their fathers, husbands or brothers.

British Association meetings, held in a different city each year, worked to raise awareness among those present that – whether industrialists, lawyers, clergy, teachers, doctors, military men or bankers – they were all engaged in the enterprise of science, a great edifice to which everyone could add a brick or two. The great men up front could suggest observations in astronomy or natural history that amateurs might make, which would be of real value in extending or testing theories. Wealthy men bought telescopes, and working men formed field clubs and went botanising or geologising,[31] stimulated by the BAAS; and its meetings were for many an agreeable, intellectual holiday, with the chance (rare in the days before radio and television) to hear and see, perhaps even meet, the heroes of science. Because the meetings were usually profitable (one in Cork being an unfortunate exception), the association was able to make grants for scientific projects. This made it a valuable patron and meant that those without institutional support could get funding.

Provincial science was thus stimulated in a variety of ways; and what might have remained an élite and metropolitan activity became a national one. This national quality, and the fact that meetings were well reported, gave the president for the year an authority as spokesman for science; rather like that the Moderator of the Church of Scotland in his domain, perhaps. The President's remarks were noticed in government circles; and sometimes he would be asked to lead a deputation to government ministers (whom, in that small world, he probably knew already) to press some case for science.

Singing from the same hymn sheet

The association set up committees 'on the state of science' in various fields, and their conclusions took up some space in its annual *Report*. These might be a mass of data, like the 326 pages of tables on earthquake phenomena presented in 1854; or they might lead to, or conclude, a debate. If there was some great issue to resolve, the meetings might function like a General Council of the Church fixing the date of Easter.

For example, in Dublin in 1835 an important question was whether chemists should use Dalton's hieroglyphic symbols for atoms of the chemical elements, where for example ⊙ meant copper and a solid black circle carbon, or the alphabetical convention of Jacob Berzelius, where B means boron, C carbon, Cu copper, and H hydrogen. To Dalton's chagrin, the meeting chose the latter, and modern chemical formulae and equations followed.[32]

In this case, Britain was somewhat belatedly agreeing to join Europe, as also nearly forty years later when there was agreement to use grams, metres and degrees Celsius rather than ounces, inches and degrees Fahrenheit; but sometimes resolutions of the BAAS could form worldwide usage. An example was Hugh Strickland's proposal for the valid naming of organisms, accepted by a committee (of which he was secretary) including Darwin, Owen and other eminent naturalists and adopted as a code by the association in 1842. This ended much of the confusion over 'synonyms' – names given by different authors to (what turned out to be) members of the same species – by favouring the oldest name and providing conventions for revisions of genera.[33] In such cases the BAAS could speak with the authority of the church scientific: the question had been fully aired and resolved, by those best qualified.

Sometimes there was heresy. The so-called Quinary system was favoured in the 1830s because of its elegant arrangement of species, genera and other groups in recurring patterns of circles at different levels. With its Trinitarian basis it had always had its critics, but it did intrigue Darwin, Huxley and Wallace, all of them seeking a pattern in nature (and eventually finding it in evolution).

It was Strickland at the Glasgow BAAS meeting in 1840, and again with the assistance of a large chart (recently found) at Cork in 1843, who is generally held to have given the 'procrustean' notion its death-blow, in a witty and formidable dismissal of its claims. He demonstrated to the satisfaction of those present[34]

> that the true affinities of organic structures branch out irregularly in all directions and that no symmetrical arrangement or numerical uniformity is discoverable in the system of nature when studied independently of preconceived theory.

Like the condemnation by Athanasius of Arius, this assault seems to have carried the assembled council with it. Quinarians were henceforward seen as having blundered, hopelessly prejudiced by their Trinitarian preconceptions, which made them see only what they were looking for. They became a kind of small sect, flourishing only in more remote places like Toronto[35] or New Zealand, to which Swainson had emigrated.

Chemist-breeding the Liebig way

Real Church Councils have on the platform their bishops and abbots, who have been consecrated, enthroned or installed in office. The scientific pundits at the BAAS – and similar bodies in the USA,[36] Australasia[37] and France – came to have similar authority but not the same standing. Although election to office signified eminence, there was still no equivalent to ordination, that set one apart and gave validity. Here again the German example was crucial. Their universities had been reformed after the Napoleonic invasion and conquest,[38] as had medicine.[39] Controversies were debated publicly in journals, and research and innovation had become more highly valued in the academic world.

Justus Liebig, after studying with Joseph Gay-Lussac in France, began training pharmacists and chemists for the PhD degree by research. His success in 'chemist-breeding'[40] and his new system of analysis for organic compounds brought his name to the fore in the scientific world. He was not unique in building up a research school,[41] but he published his students' research in the journal he controlled (it became known as *Liebigs Annalen*) and, as their doctor–father, began finding them jobs, increasingly in industry. At first, an able student might take less than a year to complete his research and defend his thesis: getting a PhD was much less formal. In Britain and the USA, Liebig became famous for his work on fertilisers and his popular writings, and he attracted English-speaking students.

Soon, a reference from Liebig was a passport to an academic post at home; his books were translated, and at meetings of the BAAS he was a welcome and honoured guest. He made the little University of Giessen world-famous and he helped to make chemistry, with its exciting combination of experiment, theory and usefulness, the leading science of the nineteenth century.

In some ways getting the PhD is akin to ordination, to joining a priesthood. The degree is conferred at a cheerful but solemn public ceremony, and it marks a change of status: my supervisor said to me 'Now we are colleagues; call me Alistair'. I had joined the clerisy. In the later nineteenth century, scientists (chemists especially) used to delight in tracing back their intellectual descent, perhaps through Liebig to Gay-Lussac to C.L. Berthollet, to the founding fathers of modern chemistry, the circle of Lavoisier – rather like an apostolic succession or the passing down of authentic Moslem traditions.

Before the PhD and its recognised status, Faraday came to the end of his apprenticeship to a bookbinder, and found a job as a master craftsman. He then joined Davy, but some years later they fell out because Davy failed to perceive that his 'apprentice' was now an independent master.[42] Engineers, like bookbinders, had a fixed term of apprenticeship; but in 1820 there was no way to become recognised as a qualified or chartered scientist. A PhD is also like the 'masterpiece' produced at the end of an apprenticeship: for the scientist is craftsman as well as 'cleric'.

We can see the PhD as the gateway to a scientific career – and Liebig has been compared to a gatekeeper – though strictly that became true only in the twentieth century, when universities all over the world began to award PhDs. From this clerisy, the 'bishops' of science emerged as academicians, Fellows of the reformed Royal Society, eminent professors and even industrialists like Lord Armstrong, delivering their messages *ex cathedra* from their university or presidential chairs.[43]

Cathedrals of science

Cathedrals are so named because they house the bishop's chair or throne, and a church scientific would need an equivalent, a place where orders could be conferred. They came in the form of museums and institutes, which were described indeed as 'cathedrals of science'. They might be in the classical style, with its suggestion of order and enlightenment, like the chemistry laboratories put up in Coimbra at the end of the eighteenth century, and the buildings of University College, London; or gothic, with its aspirations after the sublime, the style which in the English-speaking world seemed so apt for academic buildings in the nineteenth century.

The museum in Oxford, opened in time for the British Association's visit in 1860, was in John Ruskin's favourite Venetian gothic style; attached to it was the chemistry laboratory, on the model of the Abbot's Kitchen at Glastonbury – the nearest medieval equivalent. The British Museum, with its portico and great dome, was in the classical style; but when the Natural History collections were moved to South Kensington, the new building was gothic. Such museums were centres of research, with taxonomists and other specialists working in back rooms, and they were also places of education and rational amusement. Just across the road at South Kensington, Thomas Huxley – in a Renaissance-style terracotta building – lectured on zoology and got his students down to lab work rather than just description.

The message had come home to the English-speaking countries – following the crushing Prussian victory over France in 1870 – that research-based universities on German lines[44] were essential in the modern world, along with Polytechnics on the French model (or the German *Technische Hochschulen*). This lesson came home even more forcibly to France, where Adolphe Wurtz and Louis Pasteur had in different ways promoted research; but French institutions were poorly funded and highly centralised,[45] and French science seemed to be declining absolutely – not just in its ranking.

This new impetus had the further effect of at last providing posts – 'livings' one might call them – in industry[46] (full-time researchers now, as well as consultants) and in education. Schools now taught science and so did the new civic and land-grant universities. These were beginning to flourish – following in Britain the report of the 1872 Royal Commission, chaired by the Duke of Devonshire. This included prominent scientists, now part of the establishment.[47] By the time Huxley died in 1895,[48] the aspiring scientist could study chemistry, physics and biology at school, enter a university to read for a science degree, then go on to research (perhaps in Germany for a PhD). A career in science was open to anyone with talent, since patronage was now in the hands of career scientists like Huxley rather than aristocrats, and it might lead to eminence on a level with bishops.

As in the Church, overstocked with curates, it was getting on to the second rung of the academic ladder that might be difficult. Some scientists, like Crookes and Lockyer, employed assistants whose time and discoveries belonged to the boss. The Germans had invented research students, working under direction in a laboratory; and eminent professors got a lot of their research done by deputy that way. Demonstrating, tutoring and carrying the professor's bag would, once the thesis was completed, with luck lead to a position with future security.

Science had its hierarchy, less obvious than that of the Church, but very evident to those in the know; and training in a research school had much in common with training in a theological college like Cuddesdon. Wollaston was called 'the Pope' because his analyses were infallible; many since have deployed equally great influence and patronage.

Fighting the good fight: Wilberforce vs. Huxley

In this newly specialised world, it was essential to keep amateurs in their place, to ensure places for professionals. This might involve white-collar unionism, ensuring that jobs in water analysis or pollution control (the Alkali Inspectorate) went to the properly qualified; but it also involved intellectual border skirmishes. One that acquired the status of myth was the encounter between Wilberforce, Bishop of Oxford – but holding forth in the new cathedral of science there – and Huxley.

The debate was not part of the British Association's official programme: it was tacked on at the end when the journalists had knocked off. Huxley, having had a major row earlier with Owen about the gorilla's resemblance to us,[49] had intended to leave and rejoin his family but was persuaded to stay. Wilberforce was a Vice-President for the BAAS meeting, as a prominent local figure with a genuine interest in science. He delighted in the birds at Cuddesdon, and fed them; but, unlike St Francis, does not seem to have preached to them.

Descriptions of the debate[50] are not much more reliable than those of Saint George's battle with the dragon, but it began with a long pro-Darwinian speech by J.W. Draper from New York, whose *History of the Conflict between Religion and Science* was to become a classic.[51] He seems to have bored the huge audience; the eloquent Wilberforce delighted them. He could not resist ending by asking Huxley whether he was descended from an ape on his grandmother's or his grandfather's side – which may have seemed in poor taste. It perhaps riled Huxley, who retorted (variously quoted) that he would rather be descended from an ape than from one who mocked the efforts of serious investigators.

This response may have turned the tables, though it was perhaps Joseph Hooker who spoke best on the Darwinian side. At the end, both sides felt they had won (there was no vote). At any rate, Darwin's theory was taken seriously and not laughed out of court, as *Vestiges* had been. Later, this clash came to be seen as charged with deep significance. Certainly, it began Huxley's transformation into a public figure; and that in itself was important.

Perhaps more crucial was the implication of what Huxley said, that science was a specialised activity where non-specialists had no right to opinions. Herschel had made that point much earlier about astronomy, that mathematics was essential, but Huxley's was a more general (and debatable) point. It made sense in the context of academics wresting control of education from the hands of the clergy into those of clerisy, suitably qualified.

But the fact was that Wilberforce was not an ignorant or bigoted outsider; he had been coached in his arguments by the great naturalist Owen, staying with him at Cuddesdon; and the points he made, about Darwin's theory being no more than a likely story, were uncomfortably close to what Huxley had said himself not long before in a Royal Institution lecture.

If, like Karl Popper, we see science as generalisations that are falsifiable (but not falsified), then Wilberforce's criticisms were an important part of the process of testing. They were not – as Huxley seemed to be saying – improper or obscurantist interventions, or assaults upon manifest and value-free truths, or attacks on impartial truth-seekers. In fact, Wilberforce's objections were those of philosophers (and natural philosophers), expecting science to be inductive or deductive, and of many taxonomists.

Scientists are deeply conservative, suspicious of innovations and wide-ranging theories, especially if these unsettle deeply held convictions that make sense of their work. Dogma has an important place in scientific education and practice.[52] Those who classified species naturally saw species as stable, definite and real entities, not transient forms evolving all the time into something quite different. Opposition to Darwin was often scientific.

Many clergy and laymen found no difficulty in accepting an evolutionary picture of some kind: among the educated, that was not a major issue, but the question of authority and influence was. Huxley and Wilberforce, agnostic and bishop, did differ profoundly over their beliefs; but their clash may be seen as part of a campaign in which the Church Scientific captured territory from the Church of England.

A new world-view – perhaps a more materialistic one – was becoming established. The nineteenth century saw a social and institutional change, with men of science emerging as powerful figures in the educational world, clerisy displacing clergy. By 1900, science looked much more like a church than it had in 1800 – and, just as churches had seldom taken women seriously, so too the institutions of science. We shall next look at science as a bastion of machismo.

CHAPTER 12

Mastering nature

NEWTON'S *PRINCIPIA* (1687) HAS BEEN memorably compared to a 'rape manual'. This remark by a feminist seems at first sight paradoxical: anybody looking for advice on the assault, battery or conquest of women would be sorely disappointed by this daunting Latin tome on the laws of mechanics and gravity, and on planetary orbits. Newton's sex-life seems to have been meagre. But science has always been a macho business,[1] not merely because for centuries it was the province of men, but because it is concerned with mastering nature.

It is she, Mother Nature, who is raped – or at least forced into subservience – by modern science and technology, for which Newton can stand as the symbol. Francis Bacon wrote that knowledge itself is power; and power is certainly what a lot of science is concerned with. That is why governments support it – though that didn't begin to pay off until the nineteenth century. As a judge, Bacon was concerned with witch trials. An important part of his programme for experimental science was putting nature to the question: not just admiringly enjoying her company, watching and listening to her like a lover, but torturing her until she gave the required answer.

Consuming love

Where Davy, Tyndall and others may have thought they worshipped magnificent and noble nature, others might say they assaulted her – or it. To the orthodox Christian, after all, the creation was like a machine set going by the Watchmaker, rather than a goddess or an animal; and that may be have been the mainspring of the 'scientific revolution' of the seventeenth century, because machines can be fully understood.[2] Even Christians, though – who worshipped the Creator and admired His creation – have been accused of raping nature, over-exploiting the Earth, increasing and multiplying beyond safe limits and pursuing growth (or 'happiness') with indifference to the long term.

Eco-theology[3] is now thought a good thing, sustainability brings out church people in demonstrations, and ruthless greed has always been condemned by churchmen. At the same time they interpreted the first chapters of Genesis as more than a story. Apart from describing the six days of creation and the Garden of Eden, Genesis also laid down the ground rules for relations between the sexes, and the control or stewardship of the Earth, together with its animals, plants and minerals. The Book of Genesis can be (and has been) read as a licence to clear, cultivate and use to the full, rather than conserve, admire and handle with care. While humans duly gave thanks for their creation and preservation, they did not necessarily wish to preserve the world unimproved. Sailing up the coast of California, George Vancouver and his officers thought the park-like wilderness promising: it only needed to be augmented[4] by 'the neat habitations of an industrious people, to produce a scene not inferior to the most studied effects of taste in the disposal of grounds'.

The nineteenth and twentieth centuries saw the wild and the wilderness much reduced; and nostalgia – a kind of home-sickness – afflicted many. From the Romantics onwards, people longed for wide-open spaces or a home on the range; they heard, by Walden pond or Windermere, the woodlands silently pleading for peace: not just the absence of war, but contentment. Still, people were pulled two ways: after all, Henry Thoreau stayed at Walden just long enough to make a book out of it, and then returned to the city. Mastering nature, with the promise of prosperity and labour-saving devices, was quite as attractive as living in harmony with nature, going 'back to the land' and respecting God's creation.

In the engagement of science and religion, therefore, we have no simple opposition; and in both spheres the abilities and insights of women were (often still are) undervalued. Concern for the environment, unease over untrammelled technology, and concern for truth rather than simply profit, are to be found among scientists as among church leaders and members. The urgency of these concerns (the apocalyptic of the twentieth century) cuts across beliefs and they bring not peace but the sword – a two-edged sword.

Preference for an unpeopled world, disdain for vulgarity, and exalting of primitive lifestyles are not humane – they are the modern equivalents of a belief in Merrie England and its contented peasantry (with consoling political consequences for those at the top). There have always been many who looked forward, glumly or expectantly, to the end of the world – and, in a culture where scientists can be pundits or prophets, that end takes a scientific form while generating fervour like a crusade. Signs and wonders, great storms or epidemics, are interpreted as signs of the times, indicating the need for repentance, solidarity and less extravagance.

One of the great strengths of the Christian tradition is the pessimism about humans that accompanies its vision of redemption. Religion has become too often a private matter, not the place for public anxiety about global warming, radioactivity and extinctions. The church scientific and its clerisy have taken over some necessary responsibilities from Christian churches and their clergy, though often now in alliance with them as churches catch up and recover a collective vision.

Science as a spectator sport

Scientific rhetoric, like preaching, was a masculine genre, and it owed much to the theatre[5] – professing is, or should be, a performance art. This was evident in the medical dissections in the great medical schools of the Renaissance, in the anatomy theatres of Bologna, Padua or Uppsala, where the professor deserted his chair for the hands-on work of displaying the parts of the human body and their functions. In the last years of the eighteenth century, with Galvani and Volta, electricity added new excitement:[6] corpses could thereafter be made to twitch and grimace, a horrific business making spectators rush from the room with terror or sickness, even fainting.[7] Perhaps it had medical relevance when being used to galvanise lethargic teenagers into action.

Chemistry, *par excellence* an experimental science, proved more acceptable as entertainment for mixed audiences, including children: but Davy's famous inaugural lecture at the Royal Institution in 1802 was rhetorical rather than experimental. Despite coming from a radical political group in Bristol (the survivors of the Lunar Society of Birmingham),[8] Davy realised which way the wind was blowing and assured his well-heeled audience that inequality of property was vital for the advance of science and industry. He presented an entrancing vision of a country transformed by applied science, updating Bacon's hope for the restoration of Eden with his vision[9] of 'a bright day, of which we already behold the dawn'.

Davy had already, in passages that were to resonate in Mary Shelley's *Frankenstein*, spoken of the man of science interrogating nature with power, not passively as a mere scholar but actively with his own instruments: science had given him powers 'which may almost be called creative'. As well as being godlike, science was sexy: 'not contented with what is found upon the surface of the earth, he has penetrated into her bosom, and has even searched the bottom of the ocean for the purposes of allaying the restlessness of his desires, or of extending and increasing his power'.

Nevertheless, Davy deplored attempts to rip, rather than slowly lift, the veil concealing the wonderful phenomena of living nature: perhaps after all this was a strip-tease rather than a rape. Chemistry was 'sublime philosophy', though still imperfect; but the dim and uncertain alchemical twilight of discovery had given way to the steady light of truth. The man of science was the good citizen:

> The man who has been accustomed to study natural objects philosophically . . . from observing in the relations of inanimate things fitness and utility . . . will reason with deeper reverence concerning beings possessing life; and perceiving in all the phenomena of the universe the designs of a perfect intelligence, he will be averse to the turbulence and passion of hasty innovations, and will uniformly appear as the friend of tranquillity and order.

His very successful courses of lectures delivered over the next decade enthralled his hearers, confined to Britain by world war; but it became clear that for them science was a spectator sport.

A man's life

Davy was lecturing to a mixed 'Regency' audience, very unlike their children who became more prudish Victorians. Behind his rhetoric lurks the idea of carnal knowledge as an exemplar of what men of science did. Knowing meant handling, delicate or violent; it meant penetration and mastering. By contrast, deep religious knowledge was acquired through grace rather than striving or hectic activity: God spoke to Elijah not in the earthquake, whirlwind or fire, but in the still small voice; and peaceful contemplation and concentration were required of those of either gender who would be handmaids of the Lord. Religion seemed in the nineteenth century to be traditional and female, science innovative and manly.

But Davy was aware of another scientific tradition, closer to (the traditionally passive, feminine attitude of) waiting upon God. He spoke of discovery visiting the man of science; and he and his successors were aware of the importance of relaxed thinking, reverie or dreams (like Kekulé's snakes biting their tails, which led him to the benzene ring, or Mendeleyev's playing cards falling into a pattern),[10] and of the value of the prepared but unbusy mind that can take advantage of accidents. It had long been known that Nature, to be commanded, must be obeyed. Submission to the verdict of experiment, the voice of Nature, was a scientific virtue.

If Nature were the mistress of the man of science, she might be exhausting (as medieval writers saw insatiable women): harsh in the pecuniary point of view, as Davy put it to the young and starry-eyed Faraday,[11] but fascinating too and perhaps bringing stares and status; and sometimes she might be a dominatrix demanding role reversal and feminine wiles from her devotee. Science could (like religion)[12] be sublimated sex. No wonder Laetitia Barbauld, looking back in her poem on the year 1811,[13] invited her readers to:

> Call up sages whose capacious mind
> Left in its course a track of light behind;
> Point where mute crowds on Davy's lips reposed,
> And Nature's coyest secrets were disclosed.

We are back again with the strip-tease, or perhaps the seven veils.

Macho rhetoric might therefore conceal some uncertainties about whether men of science had a real job of work, or were effete intellectuals: a worrying business in the nineteenth century, with its cult of manliness. Experiment was active interrogation, to be carefully distinguished from mere observation. As the century wore on, natural history (despite Darwin) lost prestige to laboratory and museum study,[14] whereas previously field men had despised 'closet naturalists' who never left their museums.[15] There was a pecking order, an informal hierarchy of sciences, so that by the 1880s physicists looked down upon chemists as upgraded cooks, while they in turn indulgently patronised the more descriptive sciences, classifying without the numerical data of their Periodic Table.

J.J. Thomson, famous for his work on the electron, compared himself in a noteworthy passage in his autobiography to the chemist William Crookes, his predecessor as President of the Royal Society and in work on high vacua:[16]

> Crookes' success was due not only to his skill as an experimenter, but also to his powers of observation. He was very quick to observe anything abnormal and set to work to get some explanation. He tried one thing after another in the hope of increasing the effect, so as to make it easy to observe and measure; his work on the radiometer and the cathode rays are striking examples of this. In his investigations he was like an explorer in an unknown country, examining everything that seemed of interest, rather than a traveller wishing to reach some particular place, and regarding the intervening country as something to be rushed through as quickly as possible.

Thomson was that traveller, with a hypothesis (that the cathode rays were particles) to test – the twentieth-century mode. The Baconian ideal of wide-eyed and open-minded inductive inquiry (exploration, or trained and organised common sense) was replaced by the method of hypothesis, ideally quantitative, and deduction. Thomson seemed to have reduced chemistry to physics,[17] and by implication the same would happen to other sciences. Lord Kelvin, the great classical physicist, had disputed with Thomas Huxley over the age of the Earth, claiming that the laws of thermodynamics provided a tight framework, a straitjacket, within which geological theory must be fitted – and the arrogance of physicists became a feature of the first sixty years or so of the twentieth century. Physicists were sure that theirs was the fundamental science, a certainty gradually tempered by the revolutions in electronics, quantum theory and relativity.[18]

But physics, though it was and remains largely a male preserve, was not especially manly. It did not call for great physical courage, activity or endurance, or indeed for manual skills: Thomson was rather ham-handed, and Wolfgang Pauli notoriously so. Tyndall's mountaineering, with its daredevil risk-taking as well as its pantheistic meditation during solitary rambles, brought the spice of danger (and self-confidence with self-knowledge) into his laboratory life, otherwise disturbed only by public lectures. It also led him out of the comfort of the study or laboratory into research on glaciers and meteorology, work where getting cold, wet and numb was inevitable. In chemistry, it was different: the science itself was dangerous.

Experiment

Young Davy, boldly trying everything, had breathed poisonous gases such as carbon monoxide as well as the delightful nitrous oxide. Later, Faraday came as his assistant when the great man had been disabled by an explosion; and it is possible that Faraday was himself suffering from heavy-metal poisoning (his experiments involved much mercury) when he complained of loss of memory. Davy wrote that:[19]

> A steady hand and a quick eye are most useful auxiliaries; but there have been few great chemists who have preserved these advantages through life; for the business of the laboratory is often a service of danger, and the elements, like the refractory spirits of romance, though the obedient slave of the magician, yet sometimes escape the influence of his talisman and endanger his person. Both the hands and eyes of others however may be sometimes advantageously made use of.

The chemist also displayed his manliness through the handiwork and manipulation his science demanded. Apparatus could be bought by the early nineteenth century,[20] but the pioneer or discoverer, ahead of the game, would have to make his own or creatively misuse what was to hand. Chemists needed those steady hands and good lungs to blow glass, and work a blowpipe simultaneously breathing in through the nose and blowing through the mouth. They needed good senses of smell[21] and touch, and might also taste things in analysis. The hands-on business of contriving equipment might, when things went right, be followed by dancing round the laboratory in ecstatic delight.[22]

Witnesses of Davy's research were perplexed by his bursts of activity: he worked sometimes in public. For Faraday the laboratory was a peaceful place for solitary worship, with only the taciturn Sergeant Anderson present as server or acolyte. Faraday's book, *Chemical Manipulation*,[23] takes us into this world; he was proud of his father having been a blacksmith, and of his own skilled hands. The book abounds in good advice about meticulous weighing, how to bend a glass tube into a zig-zag for fractional distillation (the way he had isolated benzene), and wasting nothing. Rubber tubing was made from sheets of India-rubber, and filter papers from sheets of bibulous paper; everywhere, skilful manipulation made possible clearer experiments with smaller quantities of materials. In the interrogation of nature, it took the place of careful cross-examination.

Chemists were Wordsworthian happy warriors[24] in the popular scientific world of military metaphors, and like soldiers they risked their lives; but like craftsmen they thought with their fingers as well as their heads. Theirs was a well-rounded and manly life, but a smoky laboratory was certainly unhealthy compared with the outdoor existence of geologists. Sedgwick, investigating the Welsh rocks he named Cambrian, was breaking stones by the roadside when some ladies, taking him for a road-mender, asked him the way. They gave him a small tip – and he liked to recall that, when later he was introduced to them at dinner, he duly produced the coin. With his heavy hammer and his load of specimens, out in all weathers, the geologist really was rather like a labourer. The stone-breaking, load-carrying and scrambling required meant that a geologist had not merely to be fit, but also strong. Galton had noted how energetic and healthy his eminent scientists seemed to be;[25] Charles Darwin may have been an invalid in later life, but we should remember how active he had been earlier, on the *Beagle* and afterwards. Abundant energy seemed in those days of muscular Christianity a feature of manliness: the limp and the languid were ineffective and effeminate.

The woman and the scientist

The Victorian ideal had separate spheres for the sexes: in the man of science's household, the womenfolk were meant to cherish and support him. At the Royal Institution there is a superb photograph of Tyndall with his much younger, aristocratic wife bending solicitously over him. In Tennyson's *In Memoriam* are these chilling, haunting lines about being married to a man of science:[26]

> Their love has never passed away;
> > The days she never can forget
> > Are earnest that he loves her yet
> Whate'r the faithless people say.
>
> Her life is lone, he sits apart,
> > He loves her yet, she will not weep,
> > Tho' rapt in matters dark and deep
> He seems to slight her simple heart.
>
> He thrids the labyrinth of the mind
> > He reads the secret of the star,
> > He seems so near and yet so far,
> He looks so cold; she thinks him kind.
>
> She keeps the gift of years before,
> > A withered violet is her bliss
> > She knows not what his greatness is,
> For that, for all, she loves him more.
>
> For him she plays, to him she sings
> > Of early faith and plighted vows;
> > She knows but matters of the house,
> And he, he knows a thousand things.

> Her faith is fixed and cannot move,
>   She darkly feels him great and wise,
>   She dwells on him with faithful eyes,
> 'I cannot understand; I love'.

In fact, we know that many women actively assisted research, and understood perfectly well what the science was about. Being married to a vicar or minister was more straightforward, though loss of faith seems to have been a major risk among clergy wives too. In *Robert Elsmere*,[27] for example, the hero's wife is at a loss just like Tennyson's scientific grass-widow.

Ladies might perhaps, as Mme Lavoisier had, help in experiments. She is shown as her husband's muse in a celebrated portrait by Jacques Louis David of them in the laboratory, both clad in unsuitably sumptuous garments; and also in informal sketches of work in progress where she is keeping the records.[28] Caroline Herschel played a similarly important part in the work of her brother William, discoverer of the planet Uranus and builder of great telescopes to gauge the heavens, and later she worked independently. Ada Byron, the poet's daughter, later Countess of Lovelace, was an important mathematician, working with Babbage. She corresponded with Faraday, whose pupil she wanted to be,[29] and pumped him about his religious beliefs – rather to his alarm.

The 2001 play *Oxygen*[30] about the isolation and identification of that element, and the chemical revolution that it set off, is unusual in that the wives or women friends of the three protagonists play important parts – unexpected in the eighteenth century, but well based in history. This allows the authors to focus on the implications in science of originality, priority, communication and ethical practice in research. There is no suggestion in the play that any crucial chemical input came unattributed from the women; but the eminent Victorian historian of culture, Henry Buckle, saw a bigger role for women in science.[31]

Buckle had been educated by his mother, and he suggested in a Royal Institution Discourse in March 1858 that men had plodding inductive minds, going from facts to ideas, while women's minds were deductive, leaping from ideas to facts. He believed that too much emphasis and admiration was given to masculine inductive philosophy, which was actually contracting the minds of men of science: progress would depend upon a combination of induction and deduction, going beyond empirical laws and studying mind as well as matter, in seeking to raise the veil and penetrate into the secrets of things:

> Everything indicates that a struggle of this sort is impending, and to achieve success the imagination will have to aid the understanding more than it has done. We shall need every faculty, every resource, and every method. The intellectual peculiarities of both sexes must be combined, before we can expect to conduct to a prosperous issue that great contest between Man and Nature, of which this generation may

witness the beginning, but of which our distant posterity can hardly hope to see the end.

It was perhaps unfortunate that as an example of the feminine, deductive cast of mind, he chose Sir Isaac Newton. Somewhat similarly, in a context of nationalism rather than gender, Pierre Duhem considered that the English had broad, shallow, inductive minds, and the French deep, narrow, deductive minds: paradoxically, Napoleon was his example of an English mind, and Newton of a French one.[32]

Teach yourself science

The only science girls could expect to learn, in a world of machismo and separate spheres, would be 'domestic science' – or so we might suppose – but things seem to have been rather more complex than that. Buckle's vision of excellent science as a collaboration of men and women was not, even in his century, wholly utopian. Jane Marcet, married to a doctor who taught at Guy's Hospital in London,[33] wrote in 1807 her classic *Conversations on Chemistry*,[34] where two girls learn that science with their tutor Mrs B.

Emily and Caroline get a thoroughly up-to-date introduction to the science, from an author who had attended Davy's lectures but found they were soon forgotten unless backed up with some study:

> On attending for the first time experimental lectures, the Author found it almost impossible to derive any clear or satisfactory information from the rapid demonstrations which are usually, and perhaps necessarily, crowded into popular courses of this kind. But frequent opportunities having afterwards occurred of conversing with a friend [Dr Marcet] on the subject of chemistry, and of repeating a variety of experiments, she became better acquainted with the principles of that science, and began to feel highly interested in its pursuit. It was then that she perceived, attending the excellent lectures delivered at the Royal Institution, by the present Professor of Chemistry [Davy], the great advantage which her previous knowledge of the subject, slight as it was, gave her over others who had not enjoyed the same means of private instruction. Every fact or experiment attracted her attention, and served to explain some theory to which she was not a total stranger; and she had the gratification to find that the numerous and elegant illustrations, for which that school is so much distinguished, seldom failed to produce on her mind the effect for which they were intended.

There were no concessions to 'femininity' in these dialogues. They were not based upon cookery or needlework, but got the science across accessibly as a part of high culture.

Although she was conveying knowledge like Plato in dialogue form rather than dogmatically, Jane Marcet knew that no actual tutorial would get along as quickly as those in the book, because there the girls always say the right thing first time; but as well as discussing, they do practical experiments. 'Portable laboratories', boxes containing apparatus, could be bought; and by the early nineteenth century, chemical experiments were being done on a much smaller scale than they had been,[35] and were indeed described as suitable for the drawing room.

The success of these *Conversations* (which fired the young Faraday with enthusiasm) led to her writing series on Natural Philosophy, and on Political Economy (which delighted the young John Stuart Mill) – and to imitations, such as Elizabeth and Sarah Fitton's *Conversations on Botany*.[36] Whilst Jane Marcet's most famous readers were boys, her books (which sold well) were presumably read by just as many girls, who would thereby have picked up her enthusiasm and a lot of information – though not the 'professional' part of the science, for ladies would not be earning their living by doing chemical analyses.

Mary Somerville[37] was taken very seriously by men of science: she was a mathematician of unusual ability, capable of translating and summarising the work of Laplace. She wrote well, and she saw the big picture where others were beginning to focus sharply on their own specialism and thus be unaware of those outside it. She provided a kind of high-level popularisation, *haut vulgarisation*, so that non-specialists could keep up in a general way with their colleagues – a very important task.

Because historians of science have tended to concentrate upon originality, and on those who made some kind of important discovery, she has been, until the new biography by Kathryn Neeley, rather left out of the story, and our perception of early Victorian science is thus distorted. She could not be a Fellow of the Royal Society, but her husband was (though not what we would call a scientist) and this was a great help to her. She was an exponent of the 'scientific sublime', devoted particularly to astronomy; and for her, science, poetry and religion were all intertwined.

Science was indeed calculation and meditation. For Mary Somerville, it led to God, because it revealed the wonder, the intricacy and harmony of the whole creation. She delighted to dwell upon the way the telescope and the microscope revealed new worlds outside the realm of our unaided vision. Her readers would be awe-inspired; but there was no element of sermon about the books, because God was to be found in law, order and regularity. Miracles and special providences had no part in her world, where perfection and invariability were revealed by the progress of science, which could never be hostile to true religious belief. Her very long life (1780–1872) took her into a very different intellectual world from that in which she grew up and achieved fame; but her reputation still stood so high after her death that one of the first women's colleges in the University of Oxford was named after her. Her life was clearly exceptional.

Jobs for the girls

Mary Somerville was a very successful interpreter of science, and in that role we can also see scientific illustrators and translators. Pictures made natural history exciting and accessible,[38] and were good substitutes for actual specimens; and they were something women could do, notably in natural history. Thus Miss Drake, for example, did the illustrations for John Lindley's sumptuous *Sertum Orchidaceum* of 1838 and, with Mrs Withers, for James Bateman's enormous *Orchidaceae of Mexico and Guatemala*, 1837–42.[39] John Gould's wife Elizabeth illustrated (until her early death) his stunning bird books.[40] Anonymous women probably did the hand colouring of these plates.

Well-educated girls would not, like their brothers, usually have had to wrestle with the classics, having Latin grammar beaten into them in a male bonding exercise. Instead they studied supposedly less demanding and generally more interesting things, like drawing and modern languages. Science, down to the late seventeenth century in France and Britain and later in Germany, had been written in Latin; by 1800 hardly any of it was. The curse of Babel settled upon the learned world. Whereas men of science in Britain could rarely read German,[41] and often not much French either, educated women could make available the important work (in theology, history and philosophy as well as in science) published in these languages.

An important publisher in this field was Henry Bohn. His books, in small octavo format with small type and narrow margins, in embossed case bindings, sold very well. He was in large part responsible for the dramatic fall in book prices in the second quarter of the century, making a fortune out of it with his big sales.[42] His list included Paley's *Works*, the *Bridgewater Treatises*, and William Lawrence's 'blasphemous' physiology lectures: but he also published German works in translation, including Leopold von Ranke's *History of the Popes*, Humboldt's *Cosmos* and other writings, and Hans Christian Oersted's *The Soul in Nature*, brimming with a romantic conviction that nature is a dynamic whole[43] and that science and true religion are one.

Modern languages might be, like playing the piano, a female accomplishment rather than a key to earning a living. It is not clear whether translating could be an alternative to governessing, for the well-educated but impecunious woman needing a livelihood; but, like illustrating, it might be the basis of a respectable career. We know a lot about Mary Ann Evans, translator of the theologians Strauss and Feuerbach, because having thereby lost her faith she went to London, moved in with a married man, and became a novelist under the name of George Eliot;[44] and we know about Harriet Martineau, a formidable intellectual who edited and translated Comte and was the friend of Charles Darwin's brother Erasmus.

Humboldt's writings were translated for Bohn by Thomasina Ross and Elise Otté,[45] who 'lived wholly in the pursuit of knowledge'. Her studies included a course of physiology at Harvard as well as deep learning in Scandinavian languages and history, and she later in life wrote 'largely for scientific periodicals'. A rival translation was by Elizabeth Sabine, wife of Sir Edward Sabine, who became President of the Royal Society and shared Humboldt's interest in terrestrial magnetism.

Oersted's translators, the Horner sisters, came from a famous learned family,[46] and dedicated their translation to their friend Mathilde, Oersted's daughter. Humboldt's writings were very widely read; Oersted's, coming out after his death and representing by 1852 an old-fashioned view of science in an age of specialisms, seems to have been one of Bohn's duds, failing to catch the eye of powerful reviewers. But the role of women was crucial in sustaining the international scientific community by making classic works known across language barriers.

Victor Frankenstein's mistake

A more alarming, perhaps farsighted, view of science than Mary Somerville's was presented in Mary Shelley's *Frankenstein*. The novel outraged the reviewer in the conservative and churchy *Quarterly Review*[47] who found it a 'horrible and disgusting absurdity' of the school of Godwin – the anonymous author was in fact William Godwin's daughter, but the reviewer assumed she was male. He referred to 'passages which appal the mind and make the flesh creep', and remarked that 'our taste and our judgement alike revolt at this kind of writing'. To see the book as spooky gothick horror, a very popular genre of the day parodied in Jane Austen's *Northanger Abbey*, is understandable, and his remarks might indeed apply to some of the film versions of *Frankenstein*.

But Mary Shelley's book has passed the test of time, and unlike contemporary stories of mad monks pursuing innocent but resourceful heroines, it is still read.[48] When genetically-modified crops are referred to as 'Frankenstein vegetables', everyone gets the point, and the name is familiar even to spellcheckers. The book grew out of an evening with Percy Shelley and Byron telling ghost stories,[49] but it is full of deeper resonance and like all the best stories works on different levels – to which the reviewer might have responded, but did not.

Recent readers have picked up racism,[50] and a feminist reading is very plausible: Victor Frankenstein's is a presumptuous and unnatural male attempt to create life without women; he fails miserably to mother his offspring, and duly becomes its victim. For our purposes, it is the vision of the man of science playing God that is particularly interesting, leading on to questions of moral responsibility.

Frankenstein's enthusiasm was supposedly kindled by Professor Waldheim at the University of Ingoldstadt (in fact, a Jesuit-dominated institution associated with riots between students and soldiers, which had just been closed).[51] Waldheim's rhetoric owes much to Davy's, especially his optimism about the bright day that is coming, through the clear vision and benevolent objectives of modern men of science. God created Adam from the dust: Frankenstein creates his monster from mortuary and charnel house, galvanising him into life just as the bodies of criminals were made to twitch and grimace on the slabs of medical lecture theatres.[52]

Mary Shelley was wisely vague about the science, so the book has dated surprisingly little. God put Adam into a garden appropriate to his development; Frankenstein in panic abandoned his child in the loathsome laboratory. Seeing that Adam was lonely, God made Eve; Frankenstein, despite the monster's pleas and threats, abandoned the attempt to make a mate for him. The Garden of Eden story did not turn out very well; but the result of Frankenstein's labour was mayhem and murder. His world was ruled by ambition, fear and hatred, not by love; and the monster, initially a noble savage, took revenge upon his maker and those dear to him.

The story is set upon a iced-in ship commanded by Captain Walton on a North Polar expedition – very apposite in 1818 when the Royal Navy, with no enemy to fight any more, was being used to seek for the North-West passage around Canada[53] to the Far East. Captain Cook had, in circumnavigating the globe in high southern latitudes, met impenetrable ice and with some relief turned back, notwithstanding[54] his 'ambition not only to go farther than any one had been before, but as far as it was possible for man to go'. Walton, a less great man, is (to the horror of his crew) set upon pushing northwards, though this course looks disastrous: captains of ships, if so inclined, were able to play God. Having met Frankenstein and the pursuing monster (who makes off with his maker's corpse), Walton is shaken into responsibility, heading for home through a channel that providentially opens up in the ice.

When Mary Shelley wrote, polar exploration had been surprisingly free from utter disaster, but in 1819–22 John Franklin and his team, after reaching overland the mouth of the Coppermine River and navigating the polar sea in birch-bark canoes, had to boil and eat their boots before being rescued by Indians in northern Canada. Many years later, the two ships on Franklin's seaborne expedition were lost with all hands; and the sad fate of Captain Scott's south polar expedition is well known. We admire indomitable heroes, but perhaps we should admire more those who know when to turn back, so that they and their team live to fight another day.

Much work has gone into the search for sources for *Frankenstein*, with Genesis, Erasmus Darwin, and Shakespeare's *Tempest* among them. There are also old stories of sorcerer's apprentices, and Frankenstein is one of those: excited, indeed dazzled, by the prospect of advancing scientific knowledge, cut off from fellow human beings as he experiments hurriedly and in seclusion. He thinks in the short term and has not devoted any thought to the consequences of his actions. 'If it can be done, let's do it' is the nerd's charter, and this sort of irresponsible curiosity gave us some of the banes of twentieth-century life, as well of course as some of its delights.

It is odd that so much 'science and religion' thinking should be devoted to beliefs, rather than actions: we all know that conduct is every bit as interesting and important as thought, though philosophy of science (and the education of scientists) traditionally had little concern with ethics. In 1818, Davy had recently justified his rhetoric by inventing a safety lamp for miners: perfected in the laboratory of the Royal Institution at the smart end of London, it worked down the pits of the industrial north.

Science could thus be portrayed as benevolent: and to perceive its darker possibilities indicates how sharp Mary Shelley was. Science itself has been usefully portrayed as a Golem,[55] a monster brought into being like Frankenstein's – but by Jewish magic rather than electricity – which was a powerful, useful but dangerously dim slave. There are interesting perspectives from victims of science: from women,[56] from the history of countries like Brazil[57] and from Australian Aborigines.[58] The downside of mastering nature was starting to emerge in Mary Shelley's day.

Improving God's creation

Davy was with Stamford Raffles a founder of the London Zoo, and he hoped that exotic animals might flourish in Britain, bringing diversity to agriculture and life. The urge to 'acclimatise' foreign species, following the success of maize and potatoes in Europe, and horses in America, was prominent from the eighteenth century onwards. In the colonies, conversely, new worlds were remade in the European mode, making the land familiar.[59] Banks and those planning the First Fleet and settlement at Botany Bay took it for granted that Britons in Australia would be nourished on wheat, beef, and other familiar things rather than witchetty grubs or the meat of unfamiliar species. Then he inferred that, while the Merino sheep he had been looking after in Kew Gardens were too skinny for Britain (where lamb and mutton yields were as important as wool quality), they would flourish in Australia – where the 'botany wool' industry duly became very important.

We remember the mistakes, the grey squirrels and mink in Britain, the rabbits in Australia, the sparrows in North America; but our ancestors noticed the positive features of these exchanges, enjoying in Europe their originally exotic tomatoes, oranges, corn-on-the-cob and tobacco, while native Americans like the Apaches and Navajos rejoiced in horses and sheep, which transformed their way of life. Robert Brown, who sailed around Australia with Captain Matthew Flinders, observed how different in character and balance the flora was from that of Europe:[60] but most of his fellows continued to suppose that Europe was the norm.

Natural historians, having understood the world, had now the task of improving it. Everywhere could be made a bit more like Eden. Kew and other botanical gardens became places where foreign plants could be nurtured and then sent off to other colonies: Kew's great achievements were with quinine and rubber trees, from South America.[61] All these activities, incipient globalisation they might be called, now look less unambiguously benevolent than they did: and whereas our ancestors rejoiced in the elimination of noxious species like wolves, tidying up the creation, we look at the depleted world rather differently.

Plants need good soil and in the 1840s, the 'hungry forties', Europe was chronically short of food (as it had been half a century before during the French revolutionary wars).[62] Liebig took up where Davy had left off in agricultural chemistry, introducing inorganic fertilisers.[63] The experimental station at Rothamsted[64] (begun with profits from 'superphosphate' fertiliser made from bone ash and sulphuric acid) introduced quantitative methods, tightened up in the twentieth century by the great statistician and neo-Darwinian R.A. Fisher, to show how effective different substances really were. The battlefields of Europe were dug up for bones, and Pacific islands were stripped of the guano deposited there by seabirds over the centuries, as fertilisers became a normal part of agricultural practice. As these stocks ran down, Fritz Haber's synthesis of ammonia, meant for fertilisers but also crucial for explosives, shaped the First World War, in which Haber pioneered poison gases.

Here again, we feel concern about something that once appeared thoroughly desirable, as an expanding population was better fed. The equivocal improvement of a fallen world through chemistry[65] continued apace in the nineteenth and twentieth centuries, with plastics, synthetic dyes and explosives,[66] again making life happier in the short term but having unforeseen implications as indigo growers in India and steel workers in Europe were impoverished, and weapons of war made more deadly. By the twentieth century, much science was being done in industrial laboratories where it was more or less goal-directed, or on defence contracts from governments where secrecy was a feature, money no object, and general benevolence not very obvious – the world of Eisenhower's military–industrial complex.[67] To see science as a simple force for good was by the 1960s naïve.

Moral problems, and questions of meaning and purpose, become harder to see as an enterprise gets bigger, and sometimes they surface only in contentious projects like nuclear physics or cloning. The neutral language of protocols and procedures, and the passive voice required even of school children writing up their observations, work to distance us from what we do. Personal responsibility seems diminished. Some great scientific projects, like the NASA space programme or the Jodrell Bank radio-telescope,[68] were seen as vital to national prestige, and questions about their value in a world of limited resources seemed unpatriotic. Governments took over the task of elaborating policies for scientific and industrial research.[69]

But revulsion at the atrocities of the highly technological Nazi regime brought a new emphasis upon medical ethics, which had previously been mainly medical etiquette. In particular, the idea of informed consent has made an enormous difference within medicine and the biomedical sciences, and also beyond that in psychology, anthropology and education. Animals, often victims of mastery, began to find their defenders in the nineteenth century, with anti-vivisection movements leading to legislation on licensing: by the end of the twentieth century, the 'three Rs' – reduce, refine, replace – had become an official guide for ethics committees considering research proposals. These three demands have now to be taken seriously: the number of experiments must be minimised, the questions asked made precise, and tissue rather than living creatures must be used wherever possible.

Scientists also found themselves having to answer questions they would earlier have considered impertinent, about what long-term good might come from their experiments. To pain or inconvenience an animal by some unpleasant procedure, even to waste the time of a fellow-human through tedious and pointless questionnaires, came to be seen as ethical matters. Embarrassing or exploiting the sick, or the subjects of experiment,[70] was no longer admired; though to say it was never practised would no doubt be going too far. Deception, which had been an important feature of psychological experiments, was forbidden; and consent forms (which had been a mere formality) became a matter of taking fellow-humans seriously.

Ethical issues in research, where directors of laboratories once expected their names to be at the head of any publications, with the work of assistants and students often unacknowledged, are now regularly raised. Ethics committees have to tackle the question of how far bad science – pointless or using sloppy and inconclusive procedures – is immoral: certainly the requirement that scientists explain what they are doing to sympathetic but mildly sceptical lay people is a very good discipline.

This awareness has perhaps diminished the machismo of science, and in the twentieth century it became possible for women to play a full part[71] and for their achievements to be properly recognised; though, among the clerisy as among the clergy, there is not yet full equality. Some women, appalled by what they see as the patriarchy of the churches, have abandoned them for agnosticism or for New Age spirituality: thereby of course denying themselves possibilities of power, influence and responsibility. Medicine especially offers very popular 'alternatives' akin to these spiritualities.

It has been assumed that women should want to join fully in science and technology. If it still carries a heavy 'masculine' emphasis on mastering and penetrating, then perhaps they may feel better out of it – and in the modern world where entertainment is as prominent, if not as important, as science, they may not be foregoing much power. But certainly many do want to join in; and, in the twentieth century, scientists were also joined by men, and then women, from outside the Western, Christian tradition in which modern science matured. The science of the twenty-first century is clearly a cosmopolitan and multi-cultural activity. Finding meaning and purpose, as well as equations and laws, has got no easier: but the rich traditions opened up by an historical perspective offer much that may illuminate current perplexities.

CHAPTER 13

Meaning and purpose?

HUMANS INSTINCTIVELY LOOK FOR MEANING and purpose – in their lives, in events, in the world around them – but the search seems more difficult now than ever before. We look to science to provide answers, and then we find they do not satisfy. Douglas Adams based *The Hitch-Hiker's Guide to the Galaxy* on the premiss that our universe was created simply to provide data for a scientific experiment devised by mice in a faraway galaxy of which we know nothing. And the data? They were fed into a giant computer, which would work out the Meaning of Life, the Universe and Everything. The answer was: 42. The next stage was, of course, to work out the question.

Like every generation since Odysseus, we find life tricky. Many people need affirmation or counselling. Huxley suffered from the 'blue devils' of depression, and so do many of us – perhaps less constructively than he did. Just as the excesses of the French Revolution of 1789 were blamed on the science of the day and its dissemination by *philosophes*, so our malaise can be attributed to the scientifics (to use a Victorian term) of the recent past.

To Keats (absorbed in medicine though he was)[1] and to his friends,[2] it seemed that Newton had stolen the rainbow's meaning, the magic to which our mood could respond, and turned an enchanted world into an optical machine. There are those who feel the same about researchers in our own day and those who grumble about their broken promises (a complaint also heard from critics of the churches). Series of trumpeted 'breakthroughs' seem to lead to little of significance, and much popular science seems blinkered and deterministic.

Science can even be seen as a technocratic denial of culture (which should be one and indivisible) rather than any kind of 'second culture'.[3] If so, this denial began in the long nineteenth century, the 'Age of Science' when it came to maturity and began to make exclusive claims – but still then as part of a single high culture.[4] How did it become alienated from mainstream culture? What happened to science in the twentieth century? And where do we stand now?

The scientists dig in

By 1914 the frontiers of the various sciences were well-established, their institutions were respected and well-funded (though scientists, being human, are never satisfied and often say so). Research universities now followed the German model, with specialised institutes and laboratories catering for the various branches into which natural philosophy and natural history had diverged. Such universities were now accepted in France[5] and elsewhere in Europe, in North America, Australasia, Japan and India. The Royal Society had long had 'colonial' members[6] and by 1914 the BAAS had added Montreal and Capetown to its list of 'British' cities visited – indeed, just as the First World War broke out, it was meeting in Melbourne.[7]

Other countries adopted this model, originally German, with its annual meetings that put science firmly if sporadically on the map. Further, the French plan for an élite Polytechnic School was taken up, with more enthusiasm and a strongly industrial emphasis, in German universities[8] and the Technical High Schools that by 1914 were awarding degrees. Even in backward England there was rapid growth of 'provincial' or 'red-brick' universities, more or less committed to research, strongly inclined towards science and beginning to receive government grants for this useful work. There was also a network of polytechnics building upon the belated expansion of secondary education.[9] Students would study mathematics, physics, chemistry, biology, medicine or engineering, with increasing specialisation.

There were other research laboratories, too, especially for the booming chemical and electrical industries;[10] and there were government-sponsored institutions, the laboratory of the Government Chemist in Britain for example,[11] and the Kaiser Wilhelm Institutes in Germany, where research was separated from teaching. As countries competed industrially and militarily, and nation-states (notably Germany and Italy) were formed, nationalism became a feature of their culture and so a feature of science.[12]

Huxley's vision of a Church Scientific had indeed been realised, and its clerisy played an important part in the 'establishment' of western countries, where a scientific education was becoming increasingly common and more likely to be specialised. When conflict broke out in 1914, it soon turned into a chemists' war, where explosives and poison gases were crucial – the idea that science was simply a harmless force for good was itself exploded, and the 'military–industrial complex' that Eisenhower later perceived with alarm could already be discerned. Scientists were mobilised to develop new methods of warfare and new weapons.

In their book *The Unseen Universe*,[13] the distinguished physicists Balfour Stewart and Peter Guthrie Tait saw two sorts of people:

> We may compare the Universe to a great steamer plying between two well-known ports, and carrying two sets of passengers. The one set remain on deck and try to make out, as well as they can, the mind of

the Captain regarding the future of their voyage after they have reached the port to which they know they are all fast hastening, while the other set remain below and examine the engines. Occasionally there is much wrangling at the top of the ladder where the two sets meet, some of those who have examined the engines and the ship asserting that the passengers will all be inevitably wrecked at the next port, it being physically impossible that the good ship can carry them further. To whom those on deck reply, that they have perfect confidence in the Captain, who has informed some of those nearest him that the passengers will not be wrecked, but will be carried in safety past the port to an unknown land of felicity. And so the altercation goes on; some who have been on deck being unwilling or unable to examine the engines, and some who have examined the engines preferring to remain below.

These groups corresponded in the 1870s, that high tide of scientism, to the religious believers and the scientific naturalists, like the X-Clubbers. In the twentieth century, those involved in technological projects are no longer even 'looking at the engines'. The ordinary passengers may be as fascinated as ever by the splendid iron and brass of a paddle-steamer's machinery, but not the engineers. They are not scanning the horizon either: God's works will only divert their attention from their work, distracting them from immediate happenings but also from the big picture. Their frame does not, of course, allow for a very big picture.

Like Dr Frankenstein's, the experts' vision is as notable for what it omits as for what it foresees. It is easy to become absorbed – obsessed, even – and not see the ramifications of what one does. David Hartley in the eighteenth century noted that, in doing science, 'the Pursuit of the Pleasures of the Imagination' should not be over-indulged:[14] it 'ought to be regulated by the Precepts of Benevolence, Piety and the Moral Sense'. As we all know, it did not happen. Davy smiled at young Faraday's belief in the higher moral qualities of natural philosophers, telling him that a few years' experience would set him right.[15] Quarrels of scientists are indeed a feature of modern science,[16] as are frauds and errors.[17] It is a very human activity.

Women by 1914 were more prominent in science,[18] although it remained essentially a man's world for many years after that – and still is, in the so-called 'hard sciences'. Those early women of science in Europe and North America were joined by men and women in other countries, who had been usually 'primary producers' of raw materials, collecting data or specimens to be worked up and reasoned upon by savants in London, Paris, Munich or Boston.

'Value-free' science

Especially in Japan, the acceptance of modern science in the late nineteenth century was amazingly rapid.[19] No doubt it helped that scientific naturalism,[20] so prominent as Japan was 'opened up', was anti-Christian. Japan had nearly been made a Christian country by the Jesuits in the sixteenth century; and, in the turbulent times that led to the Shogunate being set up, thousands of Christians were martyred. The new policy of isolation was designed to keep out missionaries, who might bring in their wake the armies of the colonial powers. Christianity was presented as anti-Japanese, incompatible with patriotism and genuine culture.

Western science of the earlier nineteenth century, saturated with natural theology, would also have met with resistance: but agnostic, value-free science, in close alliance with technology, was just what the restored imperial government, opened up willy-nilly to the world by the US Navy,[21] required. It allowed Japan to maintain its independence while catching up economically and militarily with the West.

The science of the twentieth century, spread throughout the world, differed drastically from earlier science in Europe and North America: it had been detached from any religious world-view (and, often, even from any kind of spirituality); its character had become dogmatic, now that it was taught from textbooks to large classes, facing exams;[22] and it was generally indifferent to any big picture that included meaning or purpose.

Science became an enterprise for demystifying the world and bringing riches. It is now fully compatible with easygoing, vulgar materialism yet it fits just as well with belief systems, from sophisticated religions to fundamentalism, alternative medicine, nationalism, astrology – and all the other curious but evidently compelling notions that are a feature of our times. Just as the conditions of the Roman Empire in the early centuries of our era promoted the rise of Christianity, and later in the Middle East the rise of Islam, so the times were just right for scientism in the late nineteenth and the twentieth centuries.

The phenomenon of wonder

The first half of the twentieth century saw the exciting beginnings of genetics. This was generally then associated with a jerky and non-Darwinian kind of evolutionary theory, one that seemed to many people much more compatible with design and plan than the open-ended nineteenth-century version, though acceptability always varied according to time and place.[23] Paul Kammerer's work in the 1920s with salamanders indicated that animals could within a single lifetime acquire new characteristics, which their offspring could inherit; but these experiments could not be repeated. His last results, with the midwife toad, were denounced as fraudulent, and left suspicion about the rest of his work.[24] Peter Bowler has described influential attempts to reconcile science and religion in this way in Britain down to the mid century and the rise of neo-Darwinism.[25]

The last major manifestation of this came in the 1950s from the Jesuit scientist Teilhard de Chardin, whose book *The Phenomenon of Man* (1955) had been suppressed until after his death. It was therefore in essence obsolete, out of tune with what biologists in the era of DNA were saying and doing. Nevertheless, it had an astonishing if short-lived success outside that professional world because of its apparent reconciliation of evolution and purpose, presented in an attractive style. Since then, television has brought into our homes the fauna and flora of every region of the Earth, so that we can rejoice and wonder at the diversity of nature.

Watching wild animals – birds, reptiles or insects – is good for us because they take us out of ourselves: rapt, we stand and stare. They are so indifferent to us and alien from us; we need to make imaginative leaps to enter their world at all. It is wonderful to walk through a park-like landscape with giraffes, or indeed rabbits; but whereas we have mostly lost wilderness and chances of first-hand observation of larger wildlife, we have gained the opportunity for armchair travel with well-informed guides.

Cook, Banks, Humboldt, Lewis and Clark, Gosse,[26] Darwin and Wallace brought back vivid stories of remote places and their creatures, accompanied by drawings, dried pressed flowers, sad corpses preserved in spirits, or bird skins rubbed with arsenic to deter moths. Now we delight in seeing how exotic animals behave, and plants grow: the days of 'What's hit is history, but what's missed is mystery' are long past. But the objective of revealing God's existence, wisdom and benevolence that way, as Paley or Teilhard aimed to do, seems remote: all attempts to make neo-Darwinian evolution less blind, bloody and wasteful have so far failed. The much-touted 'Intelligent Design' idea seems edifying, but empty scientifically. Altruism remains hard to explain naturalistically: sex and violence rule.

Astronomers, notably Arthur Eddington and James Jeans, also featured in Bowler's story. The wonders of the starry heavens seemed even greater in the epoch of relativity theory, as faint nebulae were interpreted as island universes at vast distances, retreating from us at great speed. This gave a fresh boost to the old enterprise of astro-theology, contemplating the majesty and scale of creation, and the small size of the Earth, in a tradition going back through Newton to the Book of Job. It seemed impossible that the inexorable processes of nature might unguided (like monkeys typing Shakespeare) have thrown up so favourable a concatenation of circumstances for life as were to be found here on Earth.

Once again, the apparent conflict of chance and law seemed to be resolved by the idea of design. This was the sort of reasoning we found in Arbuthnot, on sex ratios: but by the later twentieth century, these cosmic chances could be better quantified, and they made winning the National Lottery look quite likely by comparison. We are moved by contemplation of these vast spaces and orderly, intricate movements, just as our ancestors were. It even seems that the very system has been contrived so that we not only fit it but can understand it. This is the so-called 'anthropic principle', which may not go far with anyone who has taken Hume's scepticism seriously.

And yet the Deity or First Cause to which such sublime considerations lead us is far from the loving God and Father of serious religion: we are confronted by a cleverly-contrived, inanimate, cold world. Such a world without end is not appealing. Immortality is a prominent feature of much religion, though doubtfully necessary in Christianity, for eternity can mean timelessness as easily as endless duration. In the search for immortality, we are even promised an indefinite existence in the memory of a supercomputer,[27] which seems curious but less than consoling.

One crank, two scientists, three administrators

In 1800 almost every scientific paper was written by a single author; and in 1900 many still were. The old picture persisted of the solitary genius, like Archimedes shouting 'Eureka' as the big idea suddenly dawned. But the twentieth century has been the epoch of teams, with different but complementary skills and members that come and go. The single scientist in our day is likely to be the maverick or crank, beloved by journalists, who may create panic by suggesting that MMR vaccinations cause autism or that AIDS has nothing to do with HIV.

Crookes was the last President of the Royal Society (1913–15) who was not a graduate and who did his research (with paid assistants) in his private laboratory.[28] As PhD degrees were introduced and Liebig-type graduate schools became centres for research, science was transformed. Eminent scientists became administrators. Now they managed teams of research students and did their research projects by directing others, though still expecting that their name would appear at the head of every paper from their laboratory.

The twentieth century should have been German, at any rate for science: but first the Kaiser's government in 1914, and then Hitler's from 1933, ensured that it was not. A flood of scientists, mostly Jewish, emigrated or fled Nazi Germany to the enormous benefit of scientific institutions in Britain and (especially) the USA.[29] The Second World War multiplied government funding for scientific projects; indeed, on the whole, war has been a tremendous catalyst for science. In defence, money may be no object: innovation, keeping ahead of the game, is essential. Warfare undermined the old idea that science was public knowledge, because this work was necessarily secret.

Even after the war was over, money for science continued to flow, partly because of the cold war. The 1950s and 1960s were a time of rapid change. A competitive culture emerged among researchers seeking some 'holy grail', with Nobel prizes only for the winners. Scientists can be persuaded to be surprisingly candid about this.[30] Because much work was done in industrial laboratories (notably in chemistry, pharmacy and electronics) or under industrial patronage, there was further pressure for secrecy to protect patent rights: not all science was 'public knowledge' any longer. Even so, an overview of the science of the twentieth century forms a daunting tome:[31] much of it requires technical knowledge and its very size illustrates how big science has become.

Give me a lever long enough

Meanwhile, a revolution in instruments transformed work in chemistry (unglamorous, compared to genetics or electronics, but essential).[32] At school and university in the 1950s, we aspiring chemists learnt practices that were essentially those of Liebig and the nineteenth century. We bored corks to take glass tubes connecting apparatus, because ground-glass joints were still new and expensive; we weighed with actual weights, including tiny wires that slid along the arm of the balance; we blew through wash-bottles (sometimes at each other), and down blowpipes onto samples on charcoal blocks; and we bubbled hydrogen sulphide from Kipp's Apparatus through acid or alkaline solutions.

All this, and working with Bunsen burners, retort stands, filter papers and crucibles, was very character-building. It formed still an induction into a craft and it had the macho aspects of Faraday's *Chemical Manipulation*,[33] which even after more than a century would still (had I then known of it) have been useful. Chemistry was a hands-on science, a real coming to grips with matter. Oliver Sacks' autobiography, *Uncle Tungsten*,[34] describes how the science helped him to get through the crises of growing up. It gave him skills and certainties to hold on to; it had spiritual value.

A move away from these traditional methods had begun in 1860 with the use of the spectroscope in chemical analysis, which launched Crookes' career. Chemists, though, were chary of 'physical methods' that replaced their skills and intuitions, which Primo Levi memorably described in his autobiography, *The Periodic Table*.[35] Gradually in the first half of the twentieth century, and rapidly thereafter, expensive machines using optical, infra-red, X-ray and Raman spectra, nuclear magnetic resonance, electron microscopy, and other physical phenomena came to replace those highly-valued manual skills, which had led to the infallible W.H. Wollaston being called 'the Pope' and which Faraday had so well described.

Davy had noted how Wollaston's techniques permitted analyses of far smaller quantities than had been possible in the eighteenth century. Even so, in his own analyses of pigments from ancient paintings and sculpture in Italy,[36] he had to use small samples scraped away from dark corners, which nevertheless did some harm. The new methods are quicker, more reliable, and use smaller quantities – sometimes they are completely non-destructive. Time that used to be spent on lengthy weighings, separations and analyses is now saved, so that chemists can spend more time on interesting things. Not everybody likes it, though, and it certainly makes the science less attractive to the young than when it was 'stinks', done with little thought of safety, but where thinking with the fingers and nose was required. It would be difficult today to get the audiences Watson, Beddoes or Davy attracted to series of lectures on chemistry: the subject has become technical and impenetrable,[37] and the word 'chemical' tends to arouse alarm and despondency rather than optimism and curiosity.

It is not clear where science is going. At the sharp end, specialisation and instrumentation develop apace;[38] but there will still be a vital role for the populariser, especially one who can bring home to an audience the implications for world-views. Of course, that is bound to be controversial, and may not be what leading scientists want. They hope that the public will love a slightly simplified version of the austere kind of science that appeals to the researcher, or the head of the laboratory — which is not very often the case. Popularisers know better, and we shall get their more personal view. Institutions will continue to boast of their scientific record, and take stock.[39]

What price spirituality?

So what room is there for the spiritual? We certainly still have the sublime, the very large and the very small, in a world of undiminished wonders. We can get a kick out of knowing and understanding more than we did, as we find out about the intricacies that lie behind the ordinary. Thus we do our best to retain our curiosity, to enter the kingdom like little children. And like Tyndall we can worship nature, and many do, even if we are uneasily aware with Coleridge[40]

> O Lady! we receive but what we give
> And in our life alone does Nature live

— that is, that we project perceptions on to the world rather than simply finding them there.

Science, even well pursued, will not automatically boost spirituality. For many, their spiritual life is a defiance of the humdrum, the commonplace and the predictable. For the cynics and the deluded, 'the opium of the people' needs no added science. But while experience indicates that no scientific method or discovery can establish the existence, wisdom and benevolence of God for a doubter, one or other of them can surely refine (or perhaps upset) the ideas of a believer. Thus our science can tell us what is sensible or plausible to believe, and what isn't, allowing us to discriminate between what is important in our faith, and what is mere accretion or baggage — though we are not in a region of rigorous tests. As we have seen, science need not be cold-hearted: there is room for imagination and enthusiasm.

Enthusiasm in its eighteenth-century sense, fanatical dogmatism, is another matter. Here a good dose of science should be a help, because theological posturing and quarrelling has not gone away. Sectarians might find Coleridge's dictum worth remembering, that in arguments people are generally right in what they assert, wrong in what they deny. Certainly science will do little to boost evangelical, ecclesiastical or charismatic extremism: but through natural theology it may anchor faith and indicate how God works in creation, maintaining intellectual reasonableness. The idea of original sin looks plausible once more with the inventing of computer viruses, new methods of torture and suicide bombing, where modern science and technology are abused.

In the nineteenth century, Tyndall[41] wrote scornfully about the supposed efficacy of prayer, and Galton used his statistics to investigate whether kings and queens, endlessly prayed for in churches, had lived longer than other members of their socio-economic group (like dukes and duchesses): he found, as he had expected, that they did not. It would seem that there was little more to be said, and that prayer does not work in this mechanical way. Nevertheless, there are those who are trying, with encouragement from the Templeton Foundation, to see if prayer does help cure the sick in hospital wards.

It is hard to see what outcome is in view, apart from God being mocked, when prayer is tried in a double-blind experiment. The idea is that, as in tests of a new drug, the person praying and the person prayed for would not know each other – if it worked, it would be magic rather than genuine religion. It seems, post-Holocaust and in an age of terrorism, to be an illusion that God will favour His friends in any simple way. A relationship has to be worked out between possibly complementary understandings, religious and scientific, in specific cases (Why did I get this illness?) and in a general way. The latter will still depend upon social factors, time and place. So we still find versions of Paley's arguments, and those of the Bridgewater authors, that impress some at least of us; but we are left with the questions of their contemporary critics. How far does genuine science connect with genuine religion in a serious natural theology?

Can science have laws without a lawmaker?

'Whence arises all that Order and Beauty which we see in the World?' Newton asked.[42] While physico- and astro-theology have evolved to fill new niches in the twentieth century, they can still at best indicate only that the existence of a wise and good creator is not impossible; they cannot fully answer Newton's question. The doctrine of creation, displaced in the mid-twentieth century by the neo-orthodoxy of Karl Barth,[43] has made something of a comeback, especially through green or eco-theology – but these things are focused more on our responsibilities than on the case that Paley tried to prove.

Perhaps Paley was mistaken, or his thinking was appropriate only to his time. In an age of mature science, some elements of our spirituality must be different. Physics, chemistry and biology may be the realms of explanation; religion may be that of worship and service to others; and spirituality is where delight in the particularity and beauty of things becomes apparent. David Wilson, updating Victorian analyses – some friendly, some hostile to religion – suggests in a lively new study[44] that we should see religion as the product of social evolution, with its pay-offs in the past and present. He thus boldly uses biology as a key to history and sociology, where angels might fear to tread.

Natural theology was a wonderful vehicle for popular science in its heyday; and it still works that way if we are prepared to include within its sphere a *via negativa*, with arguments against as well as for a loving creator: Richard Dawkins in there with John Polkinghorne. People do still long for scientific knowledge to be made resonant and momentous, more than mere information, even if the result is emptiness.

In our strange, eventful pilgrimage, we have met many like Newton for whom religion provided a guarantee that the world was orderly, that there were laws, and that we could (and should, as a kind of worship and proper use of our God-given capacities) find them out. Without at least a secular version of this faith in uniformity, as Balfour pointed out, it would be impossible to do science at all. Similarly, to make easier the lives of fellow-creatures, to diminish labour and suffering, has always been a major impulse towards science. But to connect particular religious beliefs or spirituality with definite scientific labours has been more difficult.

Faraday's Sandemanianism[45] no doubt reinforced his doubts about mathematics and determined his social life, but to link it to his laboratory work or to field theory is to enter very shadowy territory. The same seems to be true of others who have made it into the scientific pantheon. For Swainson and Kirby, for George Young and other 'scriptural geologists', and for Gosse,[46] however, we cannot doubt that their religious preoccupations, high-church or literalist, their glimpses of the wonderful, drove their scientific theorising.

In consequence, while their descriptive science was enduring, their names and reputations are largely forgotten and their publications are curiosities. Even Buckland, who changed his mind, is (and was)[47] usually seen as a buffoon who thought he had proved the reality of Noah's Flood. They are the losers, the heretics of science, and therefore their history should be instructive. Of course, new truths begin as heresies, as Huxley pointed out;[48] and equally it is true that modern scientists do not begin work free of any hypothesis. Nonetheless, with an eye on today's 'Creationists', we might infer that it is very unwise (as Bacon long ago indicated) to go into science with such definite assumptions – 'idols' he called them, an ironical term for monotheists – because a botch-up is the likely result.

The Bible and other sacred books or traditions, inspired though they may be,[49] are not guides to science, which requires experiment, logic and mathematics. Faraday was perhaps fortunate to get into chemistry and electricity, about which the Bible says nothing. Before contacting Davy, he had in vain tried Banks, who might have found a post in natural history for him.

Making sense of what we believe

If we try the other way, following Whewell's recommendation that science can tell us more about the God in whom we already (instinctively?) believe, and then pondering what effect scientific work has on religious beliefs, we are on more promising ground. Priestley's experiments in electricity and chemistry convinced him that matter was active and wonderful, and led him to his monism, materialism and the doctrine of the resurrection of the body. These went well with his Unitarianism, but were not a necessary part of it.

Contrariwise, Davy's later work in the same field convinced him of dualism: matter and force, particles and electric charges, were distinct, and so were bodies and immortal souls. Buckland saw, in the creations and extinctions of the deep time of geological epochs, a loving Father designing all His creatures, slowly and carefully preparing the Earth for human beings made in His image. Again, this went well with the doctrines of the church of which he was a priest, but it did not follow from them.

For Charles Darwin, this evidence of the struggle for existence revealed only a ruthless and impersonal world; but his friends and associates Gray and Wallace saw God working through it, and humans as not simply the fruit of a blind watchmaker. For Huxley, the sceptical and open-minded approach central to scientific method, and the need to fight for science as high culture, led to agnosticism. For Tyndall, imaginative reasoning – tested by observation of glaciers and weather, and coupled with German philosophy – led to awe and wonder, to worship of nature. Babbage's efforts to make clockwork computers opened for him a world of long-term programmes, developed with enormous foresight. They could even include departures from our supposed 'laws' (really mere approximations), that we called miracles. Sidgwick's work in psychical research reinforced his religious doubt; while Balfour saw in the modern physics of Crookes and J.J. Thomson evidence for pluralism. He retained his religious faith and was unworried about inconsistencies in his world-view.

In that long nineteenth century, all those we have met who looked beyond their noses built their science into their world-view; they let it illuminate their religion, or lack of it; and clearly they came to different and incompatible conclusions. This should not surprise us: on the one hand science is provisional[50] (though of course their conclusions were not propelled only by their science) and on the other hand our world-views are unstable, depending on our time, place, social position and personality.

New science makes old obsolete, as Balfour noted: and 'erroneous theories' (partial insights) have – as with 'phlogiston' and oxygen, 'caloric' and thermodynamics, and 'miasma' and epidemics – led to genuine discoveries and improvements, which then falsified them. We might wonder how timebound are religious insights. Should past ideas generally be seen, as scientists see outmoded notions, as scaffolding rather than part of the building?

Religious traditions were and are supposed to be good examples of fixity, but to Newman in the nineteenth century[51] and to David Brown at the end of the twentieth[52] they were good examples of development and change, adapted to the needs of successive generations as they interpret their inheritance. Creative inference, more or less conscious and artful in different cases, is evident in such transformations and reinterpretations. We can believe in 'eternal verities' without being so arrogant (or credulous) as to think that our understanding of them will remain the last word for all time.

For many of us the successes of modern science indicate, as they did to Baden Powell in *Essays and Reviews* way back in 1860,[53] that miracles do not happen – if by 'miracles' we mean complete breaches of the laws of nature, like the fall of heavy bodies being suspended – but that still leaves plenty of extremely unlikely events at which we can wonder. These we can fairly see as miracles, if they lead to good and make us reflect. We had a miracle in our street, when a baby fell from an upstairs window into the arms of a passing policeman; whether the child was spared for some great purpose (as Newman felt he had been from fever in Sicily) remains to be seen.

But we may wonder (or, at least, ponder) how far archaisms and impossibilities encumber religious understanding in a scientific age. Outmoded ideas can block spirituality and generosity alike. There may be more things in heaven and earth than we dream of, but to invoke them without good reason seems a cop-out. How are sublime experiences to be described? Faced now, as I write, with Michaelmas – the Feast of St Michael and All Angels – what is there to say?[54]

Angels we have heard on high

Because my wife and I thought that children should not be deceived, or invited to put their trust in non-entities, we never let our family believe in Father Christmas; and our infant Voltaires earnestly disillusioned their more credulous classmates in the kindergarten. Father Christmas with his cherubic smile, his red coat and his dear little deer seemed to us an obstacle to understanding the Christmas story.

Angels on the other hand play an important (and much depicted) role in that story. We might say that Father Christmas is just a symbol of the generous spirit that should be abroad at Christmas (and the birch twigs he carried to punish naughty children got left behind in the twentieth century). What about angels? Their importance in religious narratives makes it seem that they are more real than symbolic. Father Christmas's relations to space and time (and hence ordinary reality) are curious. Every three-year-old knows that he can be in three or four shops at once.

It was supposed that, being spiritual rather than material, angels had a different relation to space and time from ours: perhaps they took up no room, so that a myriad could dance upon a pin. Could such beings be real, and if so how would we recognise them? Perhaps by their appearance? Those of us who live near Gateshead have one answer, given the enormous steely and powerful winged being that now towers over the town's southern approach.

Traditionally, angels have feathered wings, curly hair, and may wear white clothing, or perhaps armour and a sword. Thus on the Wilton Diptych we get a wonderful medieval view of the heavenly court in blue and gold, with God surrounded by angels; those who sung to the shepherds are usually pictured, or acted in nativity plays at school, in the same way.

More strangely, seraphim have six wings, two to cover their feet, two to cover their face and two to fly; while Kirby's cherubim (before they became chubby Renaissance *putti*) were huge winged beasts like those of Babylon, intimidating indeed. If we met someone who claimed to have encountered an angel matching one of these descriptions, we would wonder perhaps if they were a danger to themselves or others; and certainly we would not expect to meet such creatures at all. Such angels have disappeared from our scientific world.

The unbeliever might say they could not disappear, because they never existed in the first place. Despite that, when people have been questioned in recent years, a majority said they believed in the existence of angels – probably at least as many people as believe in quarks or supernovae. It seems we too can meet angels, at least in dreams, either welling up from our unconscious or coming from elsewhere.

What is their function? Angels are messengers. There is a wonderful sculpture in the cathedral at Autun showing the Three Kings asleep in a bed (king-size) with their crowns on, and an angel telling them to avoid Herod and go home another way. William Blake in his painting of Jacob's Ladder,[55] shown at the Royal Academy in 1808, has a wonderful staircase spiralling up into the Sun; angelic figures go up and down, and little children who have died go upwards to be adopted and made into angels – a Swedenborgian belief often made marble in churchyards. The story of Jacob's ladder[56] indicates that earth is open to heaven, and vice versa: the terrestrial and the spiritual are constantly linked.

But who are God's messengers? We know they may not always be recognised: Balaam's donkey perceived the angel,[57] but her master was such an ass that he couldn't see it. The 'angels' of the seven churches of Asia in the Book of Revelation[58] are clergy and elders who need to get, rather than give, a message. Angels surely are other people, with a message formal or informal for us. We may entertain angels unawares, like Abraham; be kept awake tossing, turning and wrestling all night with one, like Jacob on another occasion;[59] and even exhort someone to be an angel.

We get all sorts of messages all the time, from inside or out – and we cannot prove, except by thinking and then acting, that any of them comes from a good angel – for the story of Michael and his angels defeating the Devil[60] indicates that there are bad, fallen ones as well. That may mean that we should not fret as much as we do about the sins of the flesh, to which Lucifer was not subject. We can never know beyond all doubt that a message is angelic, but the main message we get from Michael and Michaelmas is that truth is mighty and will prevail.

The great ocean of truth

'He, who begins by loving Christianity better than Truth, will proceed by loving his own Sect or Church better than Christianity, and end in loving himself better than all':[61] so Coleridge warned us. At different times, Christianity or Science has dominated European thinking. Sometimes they have exchanged insights; often, in their struggle, truth has been a casualty. We have come a long way and seen much that is sublime in science, but we have also seen the difficulties it has placed in the way of older spiritualities.

Attempts to defend some bastion of faith may well unsettle it. William Russell, *The Times* war correspondent, inspected the rusty and antiquated artillery in a Confederate redoubt, and remarked that if it were to be fired, he would feel safer in the target area than in the gun emplacement. Gosse's explosive *Omphalos*, contrary to his intentions, made the literal reading of Genesis impossible for thoughtful people; and Newman's writings, as Charles Kingsley noted, had turned many to agnosticism rather than Catholicism.[62] In our day too, 'Creationists' and atheists polarise or even create each other, like matter and anti-matter, as extremes meet. In worship the language, ritual and music take us out of ourselves, but even there the literal understanding of sacraments has caused and still causes immense difficulties.[63]

Gosse died miserable because he had not witnessed the Second Coming, the 'Rapture'; but he had often been enrapt, indeed in heaven, studying his molluscs or butterflies. It is a pity, and surely a misunderstanding, that the future tense is usually applied to what happens in the continuous present: Christ comes again, and will go on doing so.

There is plenty to reflect on, now as in the past. For one thing, in a faith refined or distilled by science, how is it that centuries of development of the Christian tradition by thinking people has led to an intellectually timid, politically conservative and sex-obsessed evangelicalism emerging as the predominant expression of Christianity? Unlike W.H. Wollaston, of course, we are fallible: but we can look forward in hope as well as humility. For, like Isaac Newton, though we are still playing on the seashore, yet the great ocean of truth lies waiting, undiscovered, before us – an ocean of spiritual truth as well as scientific truth.

Notes

1 SOMETHING GREATER THAN OURSELVES

1 F.L. Cross (ed.), *The Oxford Dictionary of the Christian Church*, London: OUP, 1957.
2 M. Mayne, *Learning to Dance*, London: Darton, Longman & Todd, 2001, p. 220.
3 G. Rowell, K. Stevenson and R. Williams (eds), *Love's Redeeming Work: the Anglican Quest for Holiness*, Oxford: OUP, 2001, p. xxxi.
4 K.A. Neeley, *Mary Somerville: Science, Illumination, and the Female Mind*, Cambridge: CUP, 2001, p. 9.
5 H.W. Piper, *The Active Universe: Pantheism and the Concept of Imagination in the English Romantic Poets*, London: Athlone, 1962, pp. 3–59.
6 J.W. Draper, *History of the Conflict between Religion and Science*, London: Henry S. King, 1875; my copy was Herbert Spencer's.
7 J.H. Brooke and G. Cantor, *Reconstructing Nature: the Engagement of Science and Religion*, Edinburgh: T. & T. Clark, 1998.
8 T. Gisborne, *The Testimony of Natural Theology to Christianity*, London: Cadell and Davies, 1818.
9 D. Norton, *A History of the English Bible as Literature*, Cambridge: CUP, 2000, pp. 387–429.
10 F.M. Turner, 'The Victorian crisis of faith and the faith that was lost', in R.J. Helmstadter and B. Lightman (eds), *Victorian Faith in Crisis: Essays on Continuity and Change in 19th-century Religious Belief*, Stanford, CA: Stanford UP, 1990, pp. 9–38.
11 P. Harrison, *The Bible, Protestantism, and the Rise of Natural Science*, Cambridge: CUP, 1998, p. 19.
12 T.H. Huxley, *Science and Hebrew Tradition: Essays*, London: Macmillan, 1901, pp. 180–1.
13 On Charles Peirce's pragmatic definition, see P.J. Croce, *Science and Religion in the era of William James: the Eclipse of Certainty, 1820–1880*, Chapel Hill, NC: Univ. of N. Carolina Press, 1995, pp. 152, 208.
14 K.A. Neeley, *Mary Somerville*, p. 196.
15 F.M. Turner, *John Henry Newman: the Challenge to Evangelical Religion*, New Haven, CT: Yale UP, 2002, pp. 322, 331.
16 A. Tennyson, *In Memoriam*, ed. S. Shatto and M. Shaw, Oxford: OUP, 1982, p. 80.
17 P.J. Croce, *Science and Religion in the Era of William James*, pp. 49–66.
18 J.R. Reid, 'T.H. Huxley and the question of morality' in A. Barr (ed.), *Thomas Henry Huxley's Place in Science and Letters: Centenary Essays*, Athens, GA: Georgia UP, 1997, pp. 31–50.
19 K. Armstrong, *The Battle for God*, London: HarperCollins, 2000.

NOTES

20 A.C. Crombie (ed.), *Scientific Change*, Oxford: OUP, 1963, pp. 347–8.
21 C.P. Snow, *The Two Cultures and the Scientific Revolution*, Cambridge: CUP, 1959, p. 3.
22 P. Bowler, *Reconciling Science and Religion: the Debate in Early 20th-century Britain*, Chicago: Chicago UP, 2001.
23 T. Sorell, *Scientism: Philosophy and the Infatuation with Science*, London: Routledge, 1991.

2 CHRISTIAN MATERIALISM

1 W. Wordsworth, *Poetry and Prose*, ed. W.M. Merchant, London: Hart-Davis, 1955, p. 367.
2 J. Harris, 'Enlightenment's Empire', *TLS*, 5066 (5 May 2000), 31.
3 K. Haakonssen (ed.), *Enlightenment and Religion: Rational Dissent in 18th-century Britain*, Cambridge: CUP, 1996.
4 M. Wheeler, *Death and the Future Life in Victorian Literature and Theology*, Cambridge: CUP, 1990.
5 J. Priestley, *The History and Present State of Electricity*, ed. R.E. Schofield, New York: Johnson, 1966, 1, p. xv. The quotation is from the 3rd edn (1775).
6 J. Priestley, *Disquisitions relating to Matter and Spirit*, 2nd edn, London: Johnson, 1782.
7 R.E. Schofield (ed.), *A Scientific Autobiography of Joseph Priestley*, Cambridge, MA: MIT, 1966.
8 I. Newton, *Opticks*, intro. I.B. Cohen [1730], New York: Dover, 1952, p. 400.
9 On 7 August 2000 there was a ceremony at Bowood, when a plaque was presented by the Royal Society of Chemistry and the American Chemical Society to mark Priestley's isolation of oxygen there.
10 J. Uglow, *The Lunar Men*, London: Faber and Faber, 2002.
11 A. Desmond and J. Moore, *Darwin*, London: Penguin, 1991, p. 5.
12 R.K. Webb, 'The Faith of 19th-century Unitarians: a Curious Incident', in R.J. Helmstadter and B. Lightman, *Victorian Faith in Crisis*, Stanford, CA: Stanford UP, 1990, pp. 126–49.
13 C. Djerrasi and R. Hoffmann, *Oxygen*, Weinheim: Wiley VCH, 2001.
14 B. Dolan, 'Conservative Politicians, Radical Philosophers and the Aerial Remedy for the Diseases of Civilization', *History of the Human Sciences*, 15 (2002), 35–54.
15 J.Z. Fullmer, *Young Humphry Davy,* Philadelphia: American Philosophical Society, 2000.
16 R.E. Schofield (ed.), *A Scientific Autobiography of Joseph Priestley*, Cambridge, MA: MIT, p. 313.
17 T.H. Levere, *Poetry Realized in Nature: Samuel Taylor Coleridge and early 19th-century science*, Cambridge: CUP, 1981.
18 T. Beddoes, *Observations on the Nature of Demonstrative Evidence*, London: Johnson, 1793, pp. 89–103.
19 M. Priestman, *Romantic Atheism*, Cambridge: CUP, 1999.
20 J.G.A. Pocock, *Barbarism and Religion*, Cambridge: CUP, 1999.
21 H.B. Carter, *Sir Joseph Banks*, London: British Museum (Natural History), 1988; R.E.R. Banks et al. (eds), *Sir Joseph Banks: a global perspective*, London: Royal Botanic Gardens, Kew, 1994.
22 J.C. Beaglehole (ed.), *The Journals of Captain James Cook, 1*, Cambridge: Hakluyt Society, 1968.
23 J. Gascoigne, *Joseph Banks and the English Enlightenment*, Cambridge: CUP, 1994; and J. Gascoigne, *Science in the Service of Empire: Joseph Banks, the British State and the*

NOTES

Uses of Science in the Age of Revolutions, Cambridge: CUP, 1998.
24 D.M. Knight in R.E.R. Banks et al. (eds), *Sir Joseph Banks: a global perspective*, London: Royal Botanic Gardens, Kew, 1994, p. 77.
25 B.S. Gower, *Scientific Method: an Historical and Logical Introduction*, London: Routledge, 1997.
26 J. Barrow, *Sketches of the Royal Society and the Royal Society Club*, London: Murray, 1849, pp. 12–53.
27 J. Golinski, *Science as Public Culture: Chemistry and Enlightenment in Britain, 1760–1820*, Cambridge: CUP, 1992.
28 J.W. Draper, *History of the Conflict between Religion and Science*, London: Henry S. King, 1875.
29 I Kings, 19: 11–12.
30 I. Jenkins and K. Sloan, *Vases and Volcanoes: Sir William Hamilton and his Collection*, London: British Museum, 1996.
31 D.M. Knight, *The Age of Science: the Scientific World-View in the Nineteenth Century*, Oxford: Blackwell, 2nd edn, 1988.
32 J.H. Brooke and G. Cantor, *Reconstructing Nature*, Edinburgh: Clarke, 1998, chapter 10.

3 WATCHMAKING

1 G.S. Kirk and J.E. Raven, *The Presocratic Philosophers: a Critical History with a Selection of Texts*, Cambridge: CUP, 1962, pp. 400–26; quotation from p. 422.
2 O. Mayr, *Authority, Liberty and Automatic Machinery in Early Modern Europe*, Baltimore: Johns Hopkins UP, 1986.
3 I.B. Cohen with R.E. Schofield (eds), *Isaac Newton's Letters and Papers on Natural Philosophy, and related documents*, Cambridge: CUP, 1958, pp. 271–394; quotation from p. 302.
4 N. Cooper (ed.), *John Ray and his Successors: the Clergyman as Biologist*, Braintree, Essex: John Ray Trust, 1999; C. Raven, *John Ray: Naturalist*, Cambridge: CUP, 1942.
5 R. Boyle, *A Free Enquiry into the Vulgarly Received Notion of Nature*, ed. E.B. Davis and M. Hunter, Cambridge: CUP, 1996; M. Hunter (ed.), *Robert Boyle Reconsidered*, Cambridge: CUP, 1994.
6 J. Ray, *The Wisdom of God Manifested in the Works of the Creation*, London: Smith, 1691, p. 34 (repr. Hildesheim: Olms, 1974).
7 C. Mather, *The Christian Philosopher: a Collection of the Best Discoveries in Nature, with Religious Improvements* [1721], intro. J.K. Piercy, Gainesville, FL: Scholar's Facsimiles, 1968.
8 N.A. Pluche, *Spectacle de la Nature: or, Nature Display'd, being Discourses of such Particulars of natural History as were thought Most Proper to Excite the Curiosity and Form the Minds of Youth*, London: Pemberton, 1733.
9 L. Euler, *Letters to a German Princess*, tr. H. Hunter, London: Murray, 1795, vol. 1, pp. 371–97.
10 See the special issue on 18th-century science lecturing, *BJHS*, 28 (1995), part 1.
11 W. Derham, *Physico-Theology: or a Demonstration of the Being and Attributes of God from His Works of Creation*, new edn, Edinburgh: Murray, 1779, p. 326.
12 W. Lilly, *Christian Astrology: Modestly Treated of in Three Books* [1647], Exeter: Regulus, 1985.
13 K. Thomas, *Religion and the Decline of Magic*, London: Weidenfeld, 1971, pp. 358–85.
14 Luke, 10: 38–42.

15 [W. Wollaston], *The Religion of Nature Delineated*, London: Lintot, 1726. Quotation from p. 59. Franklin worked on the 1725 printing.
16 J. Butler, *The Analogy of Religion, Natural and Revealed to the Constitution and Course of Nature*, new edn, London: Rivington, 1791, pp. lxiii, 380–99.
17 N. Kemp Smith (ed.), *Hume's Dialogues Concerning Natural Religion*, Oxford: OUP, 1935, quotation from pp. 260, 239, 243.
18 J.G.A. Pocock, *Barbarism and Religion*, Cambridge, CUP, 1999, vol. 2, pp. 163–257.
19 M. Eddy, MA Thesis, 'The Rhetoric and Science of William Paley's *Natural Theology*', University of Durham, 1999.
20 [H. Davy], *Salmonia*, London: Murray, 3rd edn, 1832, p. 7.
21 E. Paley, *An Account of the Life and Writings of William Paley* [1825], repr. Farnborough: Gregg, 1970, p. 110.
22 W. Paley, *The Principles of Moral and Political Philosophy*, 5th edn, Dublin: Byrne, 1793. Quotations from pp. iii, xii, 37, 44, 47; reference to Hume's *Enquiry*, pp. 10, 42.
23 W. Paley, *Horae Paulinae: or the Truth of the Scripture History of St Paul evinced by a Comparison of the Epistles which bear his Name*, London: Faulder, 1794, pp. 4, 8, 18f.
24 W. Paley, *A View of the Evidences of Christianity*, 7th edn, London: Faulder, 1800.
25 W. Paley, *Natural Theology: or Evidences of the Existence and Attributes of the Deity collected from the Appearances of Nature*, new edn, Oxford: Vincent, 1826, vol. 1, pp. 1–16.
26 R.C. Latham and W. Matthews (ed.), *The Diary of Samuel Pepys*, vol. 6, London: Bell, 1972, p. 83.
27 D. Sobel, *Longitude*, London: Fourth Estate, 1995.
28 J.C. Beaglehole (ed.), *The Journals of Captain James Cook*, vol. 2, Cambridge: Hakluyt Society, 2nd edn, 1969, p. xxxix.
29 J. Sutherland, *Victorian Fiction: Writers, Publishers, Readers*, London: Macmillan, 1995, pp. 1–27.
30 T.L. Heath (ed.), *The Works of Archimedes*, Cambridge: CUP, 1897; suppl. 1912.
31 T.L. Heath (ed.), *The Thirteen Books of Euclid's Elements*, New York: Dover, 1956.
32 W. Paley, *Natural Theology*, vol. 2, p. 152.
33 *Ibid.*, vol. 2, pp. 165f.
34 *Ibid.*, vol. 1, pp. 171, 213, 235, 256, 269.
35 *Ibid.*, vol. 2, pp. 78, 109, 151.
36 E. Darwin, *The Temple of Nature; or, the Origin of Society* [1803], repr., ed. D. King-Hele, London: Scolar, 1973; W. Paley, *Natural Theology*, vol. 2, p. 60.
37 A conference to mark the bicentenary of his death was held in Lichfield in April 2002, and the proceedings will be published; see also E. Darwin, *Cosmologia*, an edition of his poetry without the footnotes, ed. S. Harris, Sheffield: S. Harris, 2002.
38 F.E.D. Schleiermacher, *On Religion: Speeches to its Cultured Despisers*, ed. and tr. R. Crouter, Cambridge: CUP, 1988.
39 H. Chadwick (ed.), *Lessing's Theological Writings*, London: A. & C. Black, 1956.
40 J.G. Fichte, *Attempt at a Critique of all Revelation*, tr. G. Green, Cambridge: CUP, 1978.
41 F.E.D. Schleiermacher, *On Religion*, pp. 45, 98.

4 WISDOM AND BENEVOLENCE

1 I. Newton, *Opticks* [1730], ed. I.B. Cohen, New York: Dover, 1952.
2 I. Newton, *Principia Mathematica* [1687], facsimile, London: Dawson, n.d.
3 W. Blunt, *The Compleat Naturalist: a Life of Linnaeus*, London: Collins, 1971.

NOTES

4 A. Cunningham and N. Jardine (eds), *Romanticism and the Sciences*, Cambridge: CUP, 1990.
5 M.P. Crosland, *Science under Control*, Cambridge: CUP, 1992.
6 A.-H. Maehle, *Drugs on Trial: Experimental Pharmacology and Therapeutic Innovation in the 18th century*, Amsterdam: Rodopi, 1999.
7 N. Cooper (ed.), *John Ray and his Successors: the Clergyman as Biologist*, Braintree: John Ray Trust, 2000.
8 See the special issue of *BJHS*, 28 (1995), pt 1.
9 T.H. Levere and G.L'E. Turner (eds), *Discussing Chemistry and Steam: the Minutes of a Coffee House Philosophical Society, 1780–1787*, Oxford: OUP, 2002.
10 R. Bud, S. Niziol, T. Boon and A. Nahum, *Inventing the Modern World*, London: Dorling Kindersley, 2000, p. 10.
11 D. Gardner-Medwin, A. Hargreaves and E. Lazenby, *Medicine in Northumbria*, Newcastle: Pybus Society, 1993.
12 R.E. Schofield, *The Lunar Society of Birmingham*, Oxford: OUP, 1963; J. Uglow, *The Lunar Men*, London: Faber and Faber, 2002.
13 *Dictionary of Literary Biography*, Detroit: Gale, vol. 106, 1991, pp. 59–62 ('Henry Bohn').
14 I. Inkster, *The Steam Intellect Societies*, Nottingham: Nottingham UP, 1985.
15 N. Harte and J. North, *The World of University College London, 1828–1978*, London: UCL, 1978.
16 K. Hufbaur, *The Formation of the German Chemical Community, 1720–1795*, Berkeley: California UP, 1982.
17 M.P. Crosland, *In the Shadow of Lavoisier: the 'Annales de Chimie' and the Establishment of a New Science*, Faringdon: British Society for the History of Science Monograph 9, 1994.
18 Cuvier's *Essay on the Theory of the Earth* [1813] is being reprinted in a series, *The Evolution Debate*, ed. D.M. Knight, London: Routledge and Natural History Museum, 2003.
19 M. Johnson, *Bustling Intermeddler? The Life and Work of Charles James Blomfield*, Leominster: Gracewing, 2001.
20 J.R. Topham, 'Beyond the "Common Context": the Production and Reading of the Bridgewater Treatises', *Isis*, 89 (1998), 233–62.
21 W.H. Brock, *From Protyle to Proton: William Prout and the Nature of Matter, 1785–1985*, Bristol: Adam Hilger, 1985.
22 J.H. Brooke and G. Cantor, *Reconstructing Nature*, Edinburgh: Clark, 1998.
23 R. Hamblyn, *The Invention of Clouds*, London: Picador, 2001.
24 C. Bell, *Manuscript of Drawings of the Arteries*, New York: Editions Medicina Rara, n.d.
25 C. Bell, *The Hand: its Mechanism and Vital Endowments as evincing Design*, London: Pickering, 1837.
26 O. Sacks, *A Leg to Stand On*, London: Picador, 1991.
27 W. Kirby, *On the Power Wisdom and Goodness of God as manifested in the Creation of Animals and in their History Habits and Instincts*, 2nd edn, London: Pickering, 1835.
28 W. Kirby, *On the Power, Wisdom, and Goodness of God, as Manifested in the Creation of Animals, and in their History, Habits, and Instincts*, new edn, notes by T.R. Jones, London: Bohn, 1853, vol. 2, pp. 381–6; and see my 'High Church Science' in *Paradigm*, 2 (2002), 23–8.
29 W. Kirby, *On the Power . . . of God*, vol. 1, pp. 19, 21.
30 F.M. Turner, *John Henry Newman: the Challenge to Evangelical Religion*, New Haven, CT: Yale UP, 2002.

NOTES

31 M.V. Wilkes, 'Charles Babbage and his World', *Notes and Records of the Royal Society*, 56 (2002), 353–65.
32 W.W. Rouse Ball, *Cambridge Papers*, London: Macmillan, 1918, pp. 289–92.
33 K. Coburn (ed.), *Inquiring Spirit*, London: Routledge, 1951, p. 381.
34 W. Whewell, *Astronomy and General Physics, considered with reference to Natural Theology*, London: Pickering, 5th edn, 1836, p. 335.
35 M. Fisch and S. Schaffer (eds), *William Whewell; a Composite Portrait*, Oxford: OUP, 1991; R. Yeo, *Defining Science: William Whewell, Natural Knowledge and Public Debate in early Victorian Britain*, Cambridge: CUP, 1993; M. Fisch, *William Whewell, Philosopher of Science*, Oxford: OUP, 1991.
36 W. Whewell, *Astronomy and General Physics*, 1836, p. vi; and cf. p. 366f.
37 *Ibid.*, pp. 280ff.
38 [W. Whewell], *Of the Plurality of Worlds: an Essay*, London: Parker, 1853.
39 C. Babbage, *Passages from the Life of a Philosopher*, London: Longman, 1864, pp. 337–62.
40 M.B. Hall, 'The Distinguished Man of Science', in D.G. King-Hele (ed.), *John Herschel, 1792–1871: a Bicentennial Commemoration*, London: Royal Society, 1992, pp. 115–23.
41 C. Babbage, *The Ninth Bridgewater Treatise: a Fragment*, London: Murray, 2nd edn, 1838, pp. iv, xv.
42 C. Babbage, *Ninth Bridgewater Treatise*, p. 131; F. Spufford and J. Uglow ed., *Cultural Babbage: Technology, Time and Invention*, London: Faber, 1996.
43 C. Babbage, *Ninth Bridgewater Treatise*, p. 42.
44 M. Arnold, *Poetry and Prose*, ed. J. Bryson, London: Hart-Davis, 1954, pp. 144–5.
45 A. Desmond, *The Politics of Evolution: Morphology, Medicine, and Reform in Radical London*, Chicago: Chicago UP, 1989.
46 J. Scott, *A Visit to Paris in 1814*, 5th edn, 1816, and *Paris Revisited in 1815, by way of Brussels*, 4th edn, London: Longman, 1817.
47 J.R. Hofmann, *André-Marie Ampère: Enlightenment and Electrodynamics*, Cambridge: CUP, 1995, pp. 92–4, 125–36, 356–65; M.P. Crosland, *Science Under Control*: Cambridge, CUP, 1992, pp. 192–202.
48 M. Pickering, *Auguste Comte: an Intellectual Biography*, Cambridge: CUP, 1993.
49 A. Comte, *The Catechism of Positive Religion*, tr. R. Congreve, 3rd edn, London: Kegan Paul, 1891; T.R. Wright, *The Religion of Humanity: the Impact of Comtean Positivism on Victorian Britain*, Cambridge: CUP, 1986.
50 T.H. Huxley, *Collected Essays* [1894], New York: Greenwood, 1968, vol. 1, p. 156.
51 M. Wheeler, *Ruskin's God*, Cambridge: CUP, 1999, p. 25.

5 GENESIS AND GEOLOGY

1 This chapter began as a talk given at the AAAS in Los Angeles in January 1999, and then a paper discussed in a seminar at Berkeley organised by David Lindberg and Ron Numbers in April 1999. It was the basis of a lecture given at the University of Oregon, Corvallis, at the invitation of Mary-Jo Nye in November 2000, and a version was published in *Nuncius*, 15 (2000), 639–64.
2 A.R. Hall, *The Abbey Scientists*, London: Nicholson, 1966.
3 J. Hutton, *System of the Earth 1785, Theory of the Earth 1788, Observations on Granite 1794*, intro. V.A. Eyles and G.W. White, New York: Hafner, 1970.
4 J. Playfair, *Illustrations of the Huttonian Theory of the Earth*, Edinburgh: Creech, 1802.
5 M. Rudwick, 'Minerals, Strata and Fossils' in N. Jardine, J.A. Secord and E.C. Spary (eds), *Cultures of Natural History*, Cambridge: CUP, 1996, pp. 266–86.

NOTES

6. N.A. Rupke, *The Great Chain of History: William Buckland and the English School of Geology, 1819–1849*, Oxford: OUP, 1983.
7. E.O. Gordon, *The Life and Correspondence of William Buckland*, London: Murray, 1894, facing p. 32.
8. G.A. Mantell, *Petrifactions and their Teachings: or a Handbook to the Gallery of Organic Remains of the British Museum*, London: Bohn, 1851, pp. 224–313.
9. W. Buckland, *Reliquiæ Diluvianæ: or Observations on the Organic Remains Contained in Caves, Fissures, and Diluvial Gravel, and on other Geological Phenomena Attesting the Action of an Universal Deluge*, 2nd edn, London: Murray, 1824.
10. N. Cohn, *Noah's Flood: the Genesis Story in Western Thought*, New Haven, CT: Yale UP, 1996, pp. 113–26.
11. M. Rudwick, *Scenes from Deep Time*, Chicago: Chicago UP, 1992.
12. H. Davy, *Collected Works*, ed. J. Davy, London: Smith Elder, 1839–40, vol. 7, pp. 40–2.
13. *Ibid.*, vol. 7, pp. 43–4.
14. M. Wheeler, *Ruskin's God*, Cambridge: CUP, 1999, pp. 15, 182, 194.
15. L. Colley, *Britons: Forging the Nation, 1707–1837*, New Haven, CT: Yale UP, 1992, pp. 11–54.
16. J.H. Brooke, *Science and Religion: Some Historical Perspectives*, Cambridge: CUP, 1991, pp. 192–225; A. Fyfe, 'The Reception of William Paley's *Natural Theology* in the University of Cambridge, *BJHS*, 30 (1997), 321–35.
17. M.P. Crosland, *Science Under Control: the French Academy of Sciences, 1795–1914*, Cambridge: CUP, 1992.
18. J.P.R. Deleuze, *History and Description of the Royal Museum of Natural History*, Paris: Boyer, 1823.
19. H. Becher, 'Voluntary Science in 19th-century Cambridge to the 1850s', *BJHS*, 19 (1986), 57–87.
20. On Edinburgh, see J. Browne, *Charles Darwin: Voyaging*, London: Cape, 1995, pp. 36–88.
21. D.M. Knight, *Humphry Davy: Science and Power*, 2nd edn, Cambridge: CUP, 1998.
22. J. Gascoigne, *Joseph Banks and the English Enlightenment*, Cambridge: CUP, 1994, pp. 237–65; R. Banks et al. (eds), *Joseph Banks: a Global Perspective*, London: Royal Botanic Gardens, Kew, 1995.
23. N.H. Robinson and E.G. Forbes, *The Royal Society Catalogue of Portraits*, London: Royal Society, 1980, p. 18.
24. M. Rudwick, *The Great Devonian Controversy*, Chicago: Chicago UP, 1985, pp. 17–41.
25. S. Prickett, *Origins of Narrative: the Romantic Appropriation of the Bible*, Cambridge: CUP, 1996; P.C. Almond, *Adam and Eve in 17th-century Thought*, Cambridge: CUP, 1999.
26. A. Thwaite, *Glimpses of the Wonderful: the Life of Philip Henry Gosse*, London: Faber, 2002, pp. 204–27.
27. P. Butler (ed.), *Pusey Rediscovered*, London: SPCK, 1983.
28. A. Desmond and J. Moore, *Darwin*, London: Penguin, 1992, pp. 5–20.
29. B. Hilton, *The Age of Atonement: the Influence of Evangelicalism on Social and Economic Thought*, Oxford: OUP, 1988.
30. G.B. Tennyson, *Victorian Devotional Poetry: the Tractarian Mode*, Cambridge, MA: Harvard UP, 1981.
31. O. Chadwick, *The Victorian Church*, pt 1, London: A. & C. Black, 1966, pp. 167–231; and a contemporary's account, R.W. Church, *The Oxford Movement*, ed. J. Clive, Chicago: Chicago UP, 1970.

NOTES

32 F.M. Turner: *John Henry Newman: the Challenge to Evangelical Religion*, New Haven, CT: Yale UP, 2002.
33 D.J. Wilcox, *The Measure of Times Past: pre-Newtonian Chronologies and the Rhetoric of Relative Time*, Chicago: Chicago UP, 1987, wrestles with these sublime questions.
34 G. Young, *Scriptural Geology: or, an Essay on the High Antiquity attributed to the Organic Remains imbedded in Stratified Rocks*, 2nd edn, London: Simpkin Marshall, 1840.
35 T. Cooper, *Evolution, the Stone Book, and the Mosaic Record of Creation*, London: Hodder, 1878.
36 D. McCausland, *Sermons in Stones: or, Scripture Confirmed by Geology*, 2nd edn, London: Bentley, 1857.
37 E. Hitchcock, *The Religion of Geology and its Connected Sciences*, Glasgow: Collins, 1851.
38 W. Buckland, *Reliquiæ Diluvianiæ*, 2nd edn, London: Murray, pp. iv, 236.
39 Sumner's essay had been greeted enthusiastically in *The Quarterly Review*, 16 (1816–17), 37ff.
40 I. Jenkins and K. Sloan, *Vases and Volcanoes: Sir William Hamilton and his Collection*, London: British Museum, 1996.
41 C. Lyell, *Principles of Geology: Being an Attempt to Explain the Former Changes of the Earth's Surface, by Reference to Forces now in Operation*, London: Murray, 1830–33; reprint, intro. M.J.W. Rudwick, Lehre: Cramer, 1970.
42 J. Browne, *Charles Darwin: Voyaging*, London: Cape, 1995, pp. 186–90; A. Desmond and J. Moore, *Darwin*, London: Penguin, 1992, p. 108.
43 D.M. Knight, 'From Science to Wisdom: Humphry Davy's Life' in M. Shortland and R. Yeo (eds), *Telling Lives*, Cambridge: CUP, 1996, pp. 103–14.
44 H. Davy, *Consolations in Travel; or The Last Days of a Philosopher*, London: Murray, 1830, p. 135.
45 *Ibid.*, pp. 142, 145, 148, 150.
46 P.H. Gosse, *Omphalos: an Attempt to Untie the Geological Knot*, London: van Voorst, 1857; this will be reprinted in the series, *The Evolution Debate*, ed. D.M. Knight, London: Routledge and the Natural History Museum, 2003; A. Thwaite, *Glimpses of the Wonderful: the Life of Philip Henry Gosse*, London: Faber, 2002.
47 M. Johnson, *Bustling Intermeddler: The Life and Work of Charles James Blomfield*, Leominster: Gracewing, 2001, p. 35.
48 J.Z. Fullmer, *Sir Humphry Davy's Published Works*, Cambridge, MA: Harvard UP, 1969, pp. 98–100.
49 W. Buckland, *Geology and Mineralogy Considered with Reference to Natural Theology*, 2nd edn, London: Pickering, 1837, pp. 94–5. There was another edition of the book as late as 1858. The first edition is in the series *The Evolution Debate*, ed. D.M. Knight, London: Routledge and Natural History Museum, 2003.
50 W. Buckland, *Bridgewater Treatise*, London: Pickering, 1837, vol. 1, pp. 8–33; G. Wight, *Mosaic Creation Viewed in the Light of Modern Geology*, Glasgow: Maclehose, 1847, pp. 14–27.
51 W. Buckland, *Bridgewater Treatise*, p. 307.
52 L.G. Wilson (ed.), *Sir Charles Lyell's Scientific Journals on the Species Question*, New Haven, CT: Yale UP, 1970.
53 G. Cuvier, *Essay on the Theory of the Earth*, tr. R. Jameson, 5th edn, Edinburgh: Blackwood, 1827, pp. 14, 280–1; C. Lyell, *Principles of Geology*, London: Murray, vol. 1, 1830, pp. 96–9.
54 L. Agassiz, *Etudes sur les Glaciers* [1840], London: Dawson, 1966; *Studies on Glaciers*, tr. A.V. Carozzi, New York: Hafner, 1967.

55 A.C. Ramsay, 'The Old Glaciers of Switzerland and Wales' in J. Ball (ed.), *Peaks, Passes and Glaciers*, 3rd edn, London: Alpine Club, 1862, pp. 400–74; D. Bell, *Among the Rocks around Glasgow*, Glasgow: Maclehose, 1881, pp. 155–70; D. Oldroyd, 'Early Ideas about Glaciation in the English Lake District', *Annals of Science*, 56 (1999), 175–203.
56 S. Gilley, *Newman and his Age*, London: Darton and Todd, 1990.
57 J.H.C. Leach, *Sparks of Reform: the Career of Francis Jeune, 1806–1868*, Oxford: Pembroke College, 1994.
58 N. Harte and J. North, *The World of University College London, 1828–1978*, London: UCL, 1978.
59 A. Desmond, *Huxley*, 2 vols., London: Michael Joseph, 1994–7; R. Chambers, *Vestiges*, ed. J.A. Secord, Chicago: Chicago UP, 1994; B. Lightman, *The Origins of Agnosticism: Victorian Unbelief and the Limits of Knowledge*, Baltimore: Johns Hopkins UP, 1987; T.R. Wright, *The Religion of Humanity: the Impact of Comtean Positivism on Victorian Britain*, Cambridge: CUP, 1986.
60 T.H. Huxley, *Evidence as to Man's Place in Nature*, London: Williams and Norgate, 1863; reprinted in the series *The Evolution Debate*, ed. D.M. Knight. London: Routledge and Natural History Museum, 2003.
61 P. Smith (ed.), *Lord Salisbury on Politics*, Cambridge: CUP, 1972.
62 J. Morrell and A. Thackray, *Gentlemen of Science: Early Years of the British Association for the Advancement of Science*, Oxford: OUP, 1981, pp. 389–92.
63 O. Chadwick, *The Victorian Church*, London: A. & C. Black, 1966–70, pt 1, pp. 212–21; pt 2, pp. 308–27, 347–58.
64 F.M. Turner, *John Henry Newman: the Challenge to Evangelical Religion*, New Haven, CT: Yale UP, 2002, pp. 207–54, 474–526.
65 C. Daubeny, *Miscellanies*, vol. 2, Oxford: Parker, pt 4, p. 27; P. Corsi, *Science and Religion: Baden Powell and the Anglican Debate, 1800–1860*, Cambridge: CUP, 1988.
66 A.R. Hall, *The Abbey Scientists*, London: Westminster Abbey, 1966.

6 HIGH-CHURCH SCIENCE

1 [B. Jowett et al.], *Essays and Reviews*, London: Parker, 1860.
2 See Thomas Hardy's poem, 'The Respectable Burgher, on the Higher Criticism', in D. Karlin (ed.), *The Penguin Book of Victorian Verse*, London: Penguin, 1997, no. 270.
3 M. Pattison, *Memoirs of an Oxford Don*, ed. V.H.H. Green, London: Cassell, 1988.
4 E.G. Sandford (ed.), *Memoirs of Archbishop Temple*, London: Macmillan, 1906, vol. 2, pp. 419, 613: my copy was Bishop Hensley Henson's and he has annotated p. 419 'very few Clergymen will be saved' – which Temple quoted from St John Chrysostom.
5 A.J. La Vopa, *Grace, Talent, and Merit: Poor Students, Clerical Careers, and Professional Ideology in 18th-century Germany*, Cambridge: CUP, 1988, p. 198.
6 E.A. Varley, *The Last of the Prince Bishops: William Van Mildert and the High Church Movement in the early 19th century*, Cambridge: CUP, 1992, p. 153.
7 *Report of the Commissioners . . . {on} Ecclesiastical Revenues*, London: HMSO, 1835.
8 J.G.A. Pocock, *Barbarism and Religion*, 2 vols, Cambridge: CUP, 1999, dealing with Edward Gibbon, his intellectual development and relationship to Voltaire and other *philosophes*.
9 J.R. Watson, *The English Hymn; a Critical and Historical Study*, Oxford: OUP, 1999, p. 171; and for Church/Dissenter tensions, p. 338.
10 Revelation, 3: 14 –17.

NOTES

11 H.R. McAdoo, *The Spirit of Anglicanism: a Survey of Anglican Theological Method in the 17th century*, London: Black, 1965; on these various ecclesiastical terms, see F.L. Cross, *The Oxford Dictionary of the Christian Church*, Oxford: OUP, 1957.
12 Don Cupitt, 'Face to Faith', *The Guardian*, Saturday 7 July 2001.
13 R.S. Watson, *History of the Literary and Philosophical Society of Newcastle-upon-Tyne*, London: Scott, 1897; D. Gardner-Medwin, A. Hargreaves and E. Lazenby (eds), *Medicine in Northumbria*, Newcastle: Pybus, 1993; I. Inkster and J. Morrell (eds), *Metropolis and Province*, London: Hutchinson, 1983.
14 D.M. Knight, *Ideas in Chemistry: a History of the Science*, London: Athlone, 2nd edn, 1995, pp. 97ff.
15 He and other worthies are to be found in J.W. Yolton, J.V. Price and J. Stephens (eds), *The Dictionary of Eighteenth-century British Philosophers*, Bristol: Thoemmes, 1999.
16 A. Lundgren and B. Bensaude-Vincent (eds), *Communicating Chemistry: Textbooks and their Audiences*, Canton, MA: Science History, 2000.
17 On Swainson, see D.M. Knight, *Science in the Romantic Era*, Aldershot: Ashgate Variorum, 1998, pp. 197–224.
18 W. Swainson, *Taxidermy, with the Biography of Zoologists*, London: Longman, 1840, p. 347: his own biography is the longest in the book, and the frontispiece is his portrait.
19 *História, Ciências, Saúde – Manguinhos*, 8 supplemento (2001), 809–1135, is devoted to papers from a conference on the science of travellers in Brazil since 1500; see also N. Jardine, J.A. Secord and E. Spary (eds), *Cultures of Natural History*, Cambridge: CUP, 1996.
20 A.E. Gunther, *A Century of Zoology at the British Museum, through the Lives of Two Keepers, 1815–1914*, Folkestone: Dawson, 1975, pp. 28–30, 53–60.
21 C.E. Jackson, *Bird Illustrators: some Artists in early Lithography*, London: Witherby, 1975, pp. 25–31.
22 J.F.W. Herschel, *Preliminary Discourse on the Study of Natural Philosophy* [1830], repr., intro. M. Partridge, New York: Johnson, 1966.
23 M.J.S. Rudwick, *Scenes from Deep Time: early Pictorial Representations of the Prehistoric World*, Chicago: Chicago UP, 1992.
24 W. Swainson, *Preliminary Discourse on the Study of Natural History*, London: Longman, 1834.
25 J.H. Newman, *An Essay in Aid of a Grammar of Assent* [1870], ed. N. Lash, Notre Dame, IN: Notre Dame UP, 1979.
26 W. Swainson, *The Geography and Classification of Animals*, London: Longman, 1835, p. 248: Wallace's copy is at the Linnean Society – he evidently read it twice.
27 W. Jardine (ed.), *Memoirs of Hugh Edward Strickland*, London: Van Voorst, 1858, pp. 408–17.
28 J. Coggon, 'The Circular System of William Hincks', *Journal of the History of Biology* (2002), 5–42.
29 W. Swainson, *Exotic Conchology* [1834], intro. R.T. Abbott and N.F. McMillan, Princeton, NJ: Van Nostrand, 1968.
30 W. Swainson, *Preliminary Discourse*, London: Longman, 1834, pp. 367ff, 352.
31 J.R. Topham, 'Beyond the "Common Context": the Production and Reading of the Bridgewater Treatises', *Isis*, 89 (1998), 233–62; D.M. Knight, 'Genesis and Geology: a Very English Compromise', *Nuncius*, 15 (2000), 639–64.
32 W. Kirby, *On the Power, Wisdom, and Goodness of God as Manifested in the Creation of Animals, and their History, Habits and Instincts*, new edn, London: Bohn, 1853, vol. 1, pp. 18, 1, 347. The first edition of this and other Bridgewater Treatises is available (with numerous other nineteenth-century works) in microfiche from

NOTES

Chadwyck Healey (London, 2000), in the collection titled *Creation and Evolution*.
33 W. Kirby, *On the Power, Wisdom and Goodness of God*, vol. 2, pp. 179, 21, 112, 384f.
34 F.M. Turner, *John Henry Newman: the Challenge of Evangelical Religion*, New Haven: Yale UP, 2002, p. 331.
35 Based on my paper in *Zygon*, 35 (2000), 603–12.
36 S. Shapin, *The Scientific Revolution*, Chicago: Chicago UP, 1996, p. 1.
37 I.G. Barbour, *Religion in an Age of Science*, San Francisco: Harper, 1990, p. 17; C.A. Russell, *The Earth, Humanity and God*, London: UCL, 1994, p. 13.
38 R. Boyle, *The Vulgar Notion of Nature* [1686], ed. M. Hunter, Cambridge: CUP, 1996.
39 O. Mayr, *Authority, Liberty and Automatic Machinery*, Baltimore: Johns Hopkins UP, 1986, pp. 82, 102–36.
40 D. Sobel, *Longitude*, London: Fourth Estate, 1996.
41 J.H. Brooke, *Science and Religion: Some Historical Perspectives*, Cambridge: CUP, 1991, p. 192.
42 S. Schama, *Landscape and Memory*, London: Fontana, 1996.
43 A. Tennyson, *Poetical Works*, Oxford: OUP, 1953, pp. 222–3.
44 D.M. Knight, 'Presidential Address: Getting Science Across', *BJHS*, 29 (1996) 129–38, at p. 138.
45 A. Cunningham and N. Jardine (eds), *Romanticism and the Sciences*, Cambridge: CUP, 1990.
46 J.Z. Fullmer, *Young Humphry Davy: the Making of an Experimental Chemist*, Philadelphia: American Philosophical Society, 2000.
47 D.M. Knight, *Humphry Davy: Science and Power*, 2nd edn, Cambridge: CUP, 1998, pp. 36, 9.
48 J.A. Paris, *The Life of Sir Humphry Davy*, London: Colburn and Bentley, 1830, pp. 90–4.
49 M. Berman, *Social Change and Scientific Organization: the Royal Institution, 1799–1844*, London: Heinemann, 1978, p. 70.
50 H. Davy, *Collected Works*, ed. J. Davy, London: Smith Elder, 1839–40, vol. 2, p. 319; and see my 'Why is Science so Macho', *Philosophical Writings*, 14 (2000), 59–65.
51 D.M. Knight, 'From Science to Wisdom: Humphry Davy's Life' in M. Shortland and R. Yeo (eds), *Telling Lives*, Cambridge: CUP, 1996, pp. 103–14.
52 I.G. Barbour, *Religion in an Age of Science*, San Francisco: Harper, 1990, p. 48; B.M.G. Reardon, *Religion in the Age of Romanticism*, 1985, p. 5; H. Davy, *Consolations in Travel*, London: Murray, 1830, p. 219.
53 A. Desmond, *The Politics of Evolution: Morphology, Medicine and Reform in Radical London*, Chicago: Chicago UP, 1989, pp. 1–25.
54 [R. Chambers], *Vestiges of the Natural History of Creation*, ed. J. Secord, Chicago: Chicago UP, 1994.
55 A. Tennyson, *In Memoriam*, ed. S. Shatto and M. Shaw, Oxford: OUP, 1982, p. 80.
56 A.J. Rocke, *Nationalizing Science*, Cambridge, MA: MIT Press, 2001, p. 26.
57 A. Desmond and J. Moore, *Darwin*, London: Penguin, 1992, p. 5.
58 A. Desmond, *Huxley*, London: Michael Joseph, 1994–97.
59 G. Cantor, *Michael Faraday: Sandemanian and Scientist*, London: Macmillan, 1991.
60 H. Davy, *Collected Works*, London: Smith Elder, 1839–40, vol. 1, p. 185.
61 M. Shelley, *Frankenstein: or the Modern Prometheus*, Berkeley: California UP, 1994, pp. 41–3.
62 A. Tennyson, *Poetical Works*, Oxford: OUP, 1953, pp. 95–6, 525, 527.
63 Psalm 139: 9.
64 J. Glaisher, quoted in 'Introduction', *Quarterly Journal of Science*, 1 (1864), 11–12;

D.M. Knight 'Science and Culture in mid-Victorian Britain: the Reviews, and William Crookes' *Quarterly Journal of Science*', *Nuncius*, 11 (1996), 43–54.
65 J.T. Rosenberg, *Carlyle and the Burden of History*, Cambridge, MA: Harvard UP, 1985.
66 J. Smith, *Fact and Feeling: Baconian Science and the 19th-century Literary Imagination*, Madison: Wisconsin UP, 1994, pp. 11–14, 168–74.
67 J. Tyndall, *Fragments of Science: a Series of Detached Essays, Addresses, and Reviews*, London: Longman, 1899, pp. 2, 134, 197.
68 R. Barton, 'John Tyndall, Pantheist: a Rereading of the Belfast Address', *Osiris*, 2nd ser., 3 (1987), 111–34.
69 B. Lightman, *The Origins of Agnosticism: Victorian Unbelief and the Limits of Knowledge*, Baltimore: Johns Hopkins UP, 1987. Huxley and Tyndall both feature entertainingly in W.H. Mallock, *The New Republic* (1877), intro. J. Lucas, Leicester: Leicester UP, 1975.
70 F.M. Turner, *John Henry Newman; the Challenge to Evangelical Religion*, New Haven, CT: Yale UP, 2002, pp. 207–54.
71 A. von Humboldt, *Views of Nature*, tr. E.C. Otté and H.G. Bohn, London: Bohn, 1850.
72 J. Tyndall, *The Glaciers of the Alps: Mountaineering in 1861*, London: Dent, 1906, pp. 231–2, 240.
73 G.L. Geison, *The Private Science of Louis Pasteur*, Princeton, NJ: Princeton UP, 1995, pp. 234–56.
74 J. Tyndall, *The Glaciers of the Alps*, London: Dent, 1906, p. 257.

7 GOD WORKING HIS PURPOSE OUT?

1 R. Home (ed.), *Australian Science in the Making*, Melbourne: CUP, 1988; *História Ciências Saúde – Manguinhos*, 8 supplemento (2001), 809–1135.
2 A.W.N. Pugin, *Contrasts; or a Parallel between the Architecture of the 15th and 19th centuries* [1836], intro. H.R. Hitchcock, Leicester: Leicester UP, 1969.
3 J.H. Newman, *An Essay in Aid of a Grammar of Assent* [1870], intro. N. Lash, Notre Dame, IN: Notre Dame UP, 1979.
4 Ecclesiastes, 1: 9.
5 D. Hume, *Dialogues concerning Natural Religion*, ed. N. Kemp Smith, Oxford: OUP, 1935, p. 260.
6 J. Uglow, *The Lunar Men*, London: Faber, 2002.
7 E. Darwin, *The Botanic Garden* [1791], Menston: Scolar Press, 1973; D. King-Hele, *Doctor of Revolution: the Life and Genius of Erasmus Darwin*, London: Faber, 1977; D. King-Hele (ed.), *The Letters of Erasmus Darwin*, Cambridge: CUP, 1981; *Erasmus Darwin: a Life of Unequalled Achievement*, London: G. de la Mare, 1999. Dr King-Hele has resurrected Erasmus Darwin from undeserved neglect.
8 C. Powell, *The Language of Flowers*, London: Jupiter, 1977.
9 E. Darwin, *The Botanic Garden* [1791], 1973, II, lines 247–54.
10 *Ibid.*, I, p. 8 n., p. 120 n.; Notes, p. 109.
11 P.H. Barrett, D.J. Weinshank and T.T. Gottleber (eds), *A Concordance to Darwin's 'Origin of Species' first edition*, Ithaca, NY: Cornell UP, 1981. A facsimile of *On the Origin of Species*, 1st edn, will appear in the series *The Evolution Debate*, ed. D.M. Knight, London: Routledge and Natural History Museum, 2003.
12 E. Darwin, *The Temple of Nature; or, the Origin of Society* [1803], Menston: Scolar Press, 1973, p. 133.
13 Voltaire [F.A.M. Arouet], *Candide*, tr. J. Butt, London: Penguin, 1947.
14 E. Darwin, *The Temple of Nature*, 1973, p. 137.

NOTES

15 [W. Gifford (ed.)], *Poetry of the Anti-Jacobin*, London: Wright, 1799, pp. 108–41.
16 R.W. Burckhardt, *The Spirit of System: Lamarck and Evolutionary Biology*, Cambridge, MA: Harvard UP, 1977; L. Jordanova, *Lamarck*, Oxford: OUP, 1984.
17 B. de Maillet, *Telliamed: or Conversations between an Indian Philosopher and a French Missionary on the Diminution of the Sea* [1748], ed. and tr. A.V. Carrozzi: Urbana, IL: Univ. of Illinois, 1968.
18 J.B. Lamarck, *Systeme des Animaux sans Vertebres*, Paris: Deterville, 1801.
19 M. Deleuze, *History and Description of the Royal Museum of Natural History*, Paris: Royer, 1823.
20 J. Barbut, *The Genera Vermium...*, London: Sewell, 1783.
21 D.M. Knight, *Ordering the World: a History of Classifying Man*, London: Burnett, 1981.
22 J.B. Lamarck, *Philosophie Zoologique*, Paris: Dentu, 1809.
23 M.P. Crosland, *Science under Control*, Cambridge: CUP, 1992.
24 J.B. Lamarck, *Philosophie Zoologique*, Paris: Dentu, 1809, 1, pp. 6–8, xvii.
25 [R. Chambers], *Vestiges of the Natural History of Creation*, ed. J. Secord, Chicago: Chicago UP, 1994.
26 A. de Morgan, *A Budget of Paradoxes* [1872] ed. D.E. Smith, Freeport, NY: Books for Libraries, 1969, 1, pp. 15–21.
27 F.A.J.L. James (ed.), *The Correspondence of Michael Faraday*, vol. 4: 1849–55, London: IEE, 1999, p. 802.
28 J.P. Nichol, *The Architecture of the Heavens*, London: Parker, 1850: expanded from 1837 edn.
29 S.J. Gould, *Ontogeny and Phylogeny*, Cambridge, MA: Harvard UP, 1977, pp. 109–12.
30 L.A.J. Quetelet, *A Treatise on Man*, Edinburgh: W. & R. Chambers [1842], facsimile, intro. S. Diamond, Gainesville, FL: Scholar's Facsimiles and Reprints, 1969.
31 M.P. Winsor, *Starfish, Jellyfish, and the Order of Life*, New Haven, CT: Yale UP, 1976, pp. 82–97.
32 R.L. Numbers, *Creation by Natural Law: Laplace's Nebular Hypothesis in American Thought*, Seattle: Washington UP, 1977, pp. 28–35.
33 J.V. Thompson, *Zoological Researches* [1828–34], ed. A. Wheeler, London: SBNH, 1968.
34 R.B. Freeman, *The Works of Charles Darwin: an Annotated Bibliographical Handlist*, 2nd edn, Folkestone: Dawson, 1977, pp. 66–8.
35 C. Smith and M.N. Wise, *Energy and Empire: a Biographical Study of Lord Kelvin*, Cambridge: CUP, 1989.
36 M.J. Nye, *Before Big Science*, New York: Twayne, 1996.
37 J.D. Burchfield, *Lord Kelvin and the Age of the Earth*, New York: Science History, 1975.
38 H.G. Wells, *The Time Machine* [1895], ed. J. Lawton, London: Everyman, 1995.
39 *Ibid.*, p. 74.
40 P. Bowler, *The Eclipse of Darwinism: Anti-Darwinian Theories in the Decades around 1900*, Baltimore: Johns Hopkins UP, 1983.
41 H. Drummond, *The Ascent of Man*, London: Hodder and Stoughton, 1894, pp. 52, 442–4.
42 P. Bowler, 'Charles Raven and the new natural theology' in N. Cooper (ed.), *John Ray and his Successors: the Clergyman as Biologist*, Braintree, Essex: John Ray Trust, 2000, pp. 215–24.

8 LAY SERMONS

1. D.M. Knight, 'Travels and science in Brazil', *História Ciências Saúde Manguinhos*, 8 (2001), 809–22.
2. P. White, *Thomas Huxley: Making the 'Man of Science'*, Cambridge: CUP, 2003, pp. 67–99.
3. A. Thwaite, *Glimpses of the Wonderful: the Life of Philip Henry Gosse*, London: Faber, 2002, pp. 206–7.
4. U. Goodenough, 'Exploring resources of naturalism', *Zygon*, 35 (2000), 561–6, at pp. 561–2.
5. C.M. Davies, *Unorthodox London* [1875], New York: Kelley, 1969, pp. 114–15.
6. R. Holmes, *Coleridge: Darker Reflections*, London: Harper Collins, 1998, pp. 107–44, 439–49; S.T. Coleridge, *Lay Sermons* [1816–17], ed. R.J. White, London: Routledge, 1972, p. 18.
7. S.T. Coleridge, *Lay Sermons*, 1972, pp. 33–4, 43.
8. *Ibid.*, p. 171.
9. R. Holmes, *Coleridge: Darker Reflections*, London: Harper Collins, 1998, pp. 107–8.
10. K. Coburn, *Inquiring Spirit: a New Presentation of Coleridge from his published and unpublished Prose Writings*, London: Routledge, 1951, p. 411.
11. T.H. Huxley, *Lay Sermons, Addresses and Reviews*, London: Macmillan, 6th edn, 1877, pp. 176–7.
12. This discussion follows closely my 'Getting science across', *BJHS*, 29 (1996), 129–38, and 'Scientific lectures: a history of performance', *Interdisciplinary Science Reviews*, 27 (2002), 217–24.
13. G. Berkeley, 'Siris' in *Works*, 3 vols, London: Priestley, 1820, III, pp. 259–418.
14. [R. Chambers], *Vestiges of the Natural History of Creation* [1844], ed. J. Secord, Chicago: Chicago UP, 1994. Its anonymous author was referred to as 'Mr Vestiges'.
15. L. Huxley, *Life and Letters of Sir Joseph Dalton Hooker*, London: Murray, 1918, I, p. 451.
16. Quoted in F.M. Turner, *John Henry Newman: the Challenge to Evangelical Religion*, New Haven, CT: Yale UP, 2002, p. 424.
17. G.S. Haight, *George Eliot: a Biography*, Oxford: OUP, 1968.
18. A. Barr (ed.), *Thomas Henry Huxley's Place in Science and Letters*, Athens, GA: University of Georgia Press, 1997, p. 54.
19. A. Desmond, *Huxley: the Devil's Disciple*, London: Michael Joseph, 1994.
20. J.A. Paris, *The Life of Sir Humphry Davy*, London: Colburn and Bentley, 1831, p. 89.
21. H. Davy, *Consolations in Travel*, London: Murray, 1830, p. 89.
22. T.H. Huxley, *Lay Sermons*, London: Macmillan, 1877, p. 77 [original lecture 1854].
23. J. Golinski, *Science as Public Culture: Chemistry and Enlightenment in Britain, 1760–1820*, Cambridge: CUP, 1992, pp. 188–235.
24. D.M. Knight, *Science in the Romantic Era*, Aldershot: Ashgate Variorum, 1998, pp. 298, 303.
25. H. Davy, *Collected Works*, 9 vols., London: Smith Elder, 1839–40, VIII, pp. 313–14. This has been repr., intro. D.M. Knight, Bristol: Thoemmes, 2001.
26. W. Shakespeare, *The Tempest*, IV.i.151.
27. [R. Whately], *Historic Doubts relative to Napoleon Bonaparte* [1819], 7th edn, London: Fellowes, 1841.
28. D.M. Knight, *Humphry Davy: Science and Power*, 2nd edn, Cambridge: CUP, 1998, p. 144.
29. [Anon. review], 'Comparative anatomy and classification', *Quarterly Journal of Science*, 1 (1864), 544.

NOTES

30 T.H. Huxley, 'Species and races and their origin', *Proceedings of the Royal Institution*, 3 (1858–62), 195–200; A. Desmond, *Huxley: the Devil's Disciple*, London: Michael Joseph, 1994, pp. 267–70.
31 J. Tyndall, 'Alpine sculpture' and 'The scientific use of the imagination', *Fragments of Science*, London: Longman, 1899, I, pp. 229–52, II, pp. 101–34.
32 N. Cooper (ed.), *John Ray and his Successors: the Clergyman as Biologist*, Braintree: John Ray Trust, 1999; and specifically W.S. Symonds, *Old Bones; or, Notes for Young Naturalists*, London: Hardwicke, 1861.
33 J.C. Greene, *Science, Ideology, and World View*, Berkeley: California UP, 1981.
34 T.H. Huxley, 'Species and races . . .', *Proc. Royal Institution*, 3 (1858–62), 199, 200.
35 J.P. Fordresher (ed.), *A Variorum Edition of Tennyson's Idylls of the King*, New York, 1973, p. 962.
36 J.W. Draper, *History of the Conflict between Religion and Science*, London: King, 1975. The copy I have used was Herbert Spencer's, but he has not annotated it.
37 T.H. Huxley, 'The coming of age of *The Origin of Species*', *Proceedings of the Royal Institution*, 9 (1879–81), 361–8; the text is Matthew 21: 42.
38 L. Pyenson and S. Sheets-Pyenson, *Servants of Nature: a History of Scientific Institutions, Enterprises and Sensibilities*, London: HarperCollins, 1999, pp. 125–49.
39 E.R. Lankester, under 'Huxley' in R.B. Freeman, *Charles Darwin: a Companion*, Folkestone: Dawson, 1978, p. 170.
40 T.H. Huxley, *Lay Sermons*, p. 199.
41 *Ibid.*, p. 201.
42 M. Foster, in O. Lodge (ed.), *Huxley Memorial Lectures*, Birmingham, 1914, p. 36.
43 See A. Pyle (ed.), *Agnosticism: Contemporary Responses to Spencer and Huxley*, Bristol: Thoemmes, 1995, reprints 19th-century essays and reviews.
44 T.H. Huxley, *Collected Essays* [1894], reprinted New York: Greenwood, 1968, vol. 1, p. 156; see also A.P. Barr (ed.), *Thomas Henry Huxley's Place in Science and Letters*, Athens, GA: Georgia UP, 1997, p. 56.
45 Someone collected and bound into a volume, which I now have, a number of these lectures, which print (where there is room) the rules, the committee, and the back numbers available.
46 I. MacKillop, *The British Ethical Societies*, Cambridge: CUP, 1986.
47 N. Longmate, *King Cholera: the Biography of a Disease*, London: Hamish Hamilton, 1966.
48 J. Snow, *Snow on Cholera*, New York: Hafner, 1965; J. Pickstone, 'Medicine, society and the state' in R. Porter (ed.), *The Cambridge Illustrated History of Medicine*, Cambridge: CUP, 1996, pp. 304–41.
49 A.P. Stewart and E. Jenkins, *The Medical and Legal Aspects of Sanitary Reform* [1867], ed. M.W. Flinn, Leicester: Leicester UP, 1969.
50 Registrar General, *Annual Summary of Births, Deaths and Causes of Death in London and other large Cities*, London: HMSO, 1874, p. v.
51 E.R. Robson, *School Architecture* [1874], ed. M. Seaborne, Leicester: Leicester UP, 1972, pp. 163f, 197.
52 J.P. Kay, *The Moral and Physical Condition of the Working Classes Employed in the Cotton Manufacture in Manchester* [1832], repr., ed. E.L. Burney, Manchester: Morten, 1969.
53 F. Nightingale, *Notes on Nursing* [1859], repr., London: Duckworth, 1970.
54 A. Desmond, *Huxley: Evolution's High Priest*, London: Michael Joseph, 1997, p. 259.
55 W. Lawrence, *Lectures on Physiology, Zoology, and the Natural History of Man, delivered at the Royal College of Surgeons*, London: Benbow, 1822, pp. 1–14.
56 T. Fulford (ed.), *Romanticism and Science, 1773–1833*, London: Routledge, 2002,

vol. 5, pp. 45–83, reprints documents by Abernethy and Lawrence.
57 T.H. Huxley, *Science and Education*, intro. C. Winnick, New York: Citadel, 1964, p. 77.

9 KNOWLEDGE AND FAITH

1 A.R. Hall, *The Abbey Scientists*, London: Nicholson, 1966.
2 J.F.W. Herschel, *Essays from the Edinburgh and Quarterly Reviews, with Addresses and Other Pieces*, London: Longman, 1857, p. 737.
3 W. Shakespeare, *A Midsummer Night's Dream*, V.i.4–8.
4 R. Helmstadter and B. Lightman (eds), *Victorian Faith in Crisis: Essays on Continuity and Change in 19th-century Religious Belief*, Stanford, CA: Stanford UP, 1990, esp. F.M. Turner's chapter, pp. 9–38.
5 F.M. Turner, *John Henry Newman: the Challenge to Evangelical Religion*, New Haven, CT: Yale UP, 2002, pp. 513, 564.
6 G.C. Lewis, *An Essay on the Influence of Authority in Matters of Opinion*, London: Parker, 1849.
7 A. Pyle (ed.), *Agnosticism: Contemporary Responses to Spencer and Huxley*, Bristol: Thoemmes, 1995, p. 44
8 T. Sorell, *Scientism: Philosophy and the Infatuation with Science*, London: Routledge, 1991.
9 M. Hawkins, *Social Darwinism in European and American Thought, 1860–1945*, Cambridge: CUP, 1997, draws attention (p. 144) to Ernst Haeckel's pantheistic monism.
10 A. Comte, *The Positive Philosophy*, ed. and tr. H. Martineau, London: Chapman, 1853, vol. 1, p. 2. Harriet Martineau was a close friend of Charles Darwin's brother Erasmus. See also T.R. Wright, *The Religion of Humanity: the Impact of Comtean Positivism on Victorian Britain*, Cambridge: CUP, 1986.
11 See my chapter in A.P. Barr (ed.), *Thomas Henry Huxley's Place in Science & Letters: Centenary Essays*, Athens, GA: Georgia UP, 1997, pp. 51–66, esp. pp. 53–7.
12 See my paper 'A.J. Balfour: scientism and scepticism', *Durham University Journal*, 87 (1995), 23–30; repr. in D.M. Knight, *Science in the Romantic Era*, Aldershot: Ashgate Variorum, 1998, pp. 325–39.
13 P. Smith (ed.), *Lord Salisbury on Politics: a Selection of his Articles in the Quarterly Review, 1860–1883*, Cambridge: CUP, 1972.
14 H.L. Mansel, *The Limits of Religious Thought: Bampton Lectures*, 4th edn, London: Murray, 1859; B. Lightman, *The Origins of Agnosticism: Victorian Unbelief and the Limits of Knowledge*, Baltimore: Johns Hopkins UP, 1987, pp. 32–67.
15 A.J. Balfour, *A Defence of Philosophic Doubt*, London: Macmillan, 1879, p. 293; see also pp. 86, 103–8.
16 Cf. M. Midgley, *Science as Salvation: a Modern Myth and its Meaning*, London: Routledge, 1992.
17 A.J. Balfour, *The Foundations of Belief: being Notes Introductory to the Study of Theology*, 2nd edn, London: Longman, 1895, p. 31.
18 H.G. Wells, *The Time Machine* [1895], ed. J. Lawton, London: Everyman, 1995.
19 P. Metcalf, *James Knowles: Victorian Editor and Architect*, Oxford: OUP, 1980, pp. 197ff.
20 R. Barton, 'Huxley, Lubbock and half a dozen others: professionals and gentlemen in the formation of the X Club, 1851–64', *Isis*, 89 (1998), 410–44; Ruth Barton is writing a book on the X Club.
21 W. Ward (ed.), *Papers read before the Synthetic Society, 1896–1908*, London: Synthetic Society, 1909.

NOTES

22 Lord Kelvin [W. Thomson], 'Nineteenth century clouds over the dynamical theory of heat and light', *Proceedings of the Royal Institution*, 16 (1899–1901), 363–97.
23 M.J. Nye, *Before Big Science*, New York: Twayne, 1996, chapter 6.
24 A. Schuster, *The Progress of Physics during Thirty-three Years, 1875–1908*, Cambridge: CUP, 1911; J.J. Thomson, *Recollections and Reflections*, London: Bell, 1936.
25 G.G. Stokes, *Mathematical and Physical Papers*, vol. 2, Cambridge: CUP, 1883, p. 97.
26 A.J. Balfour, 'Presidential address', *Report of the British Association*, Cambridge, 1904, pp. 3–14; see esp. pp. 8, 9 and 14.
27 In a newspaper cutting inserted in my copy of Lord Rayleigh, *Lord Balfour in his Relation to Science*, Cambridge: CUP, 1930. See p. 30 for Balfour as a would-be scientist.
28 Cf. T.S. Kuhn, 'The function of dogma in scientific research', in A.C. Crombie (ed.), *Scientific Change*, Oxford: OUP, 1963, pp. 347–69.
29 See my paper 'Observation, experiment, theory – and the spirits', *Durham University Journal*, 83 (1991), 55–8; repr. in D.M. Knight, *Science in the Romantic Era*, Aldershot: Ashgate, 1998, pp. 317–24.
30 A. and E.M. S[idgwick], *Henry Sidgwick: a Memoir*, London: Macmillan, 1906, pp. 598ff.
31 K. von Reichenbach, *Researches on Magnetism, Electricity, Heat, Light, Crystallization, and Chemical Attraction, in their Relations to the Vital Force*, tr. W. Gregory, London: Taylor, Walton & Maberly, 1850.
32 C.M. Davies, *Unorthodox London*, London: Tinsley, 3rd edn, 1875, pp. 166–74, 302–29.
33 B. Stewart and P.G. Tait, *The Unseen Universe: or Physical Speculations on a Future State*, 6th edn, London: Macmillan, 1876 [the first was 1875]; and see their *Paradoxical Philosophy: a Sequel to the Unseen Universe*, London: Macmillan, 1878.
34 J. Oppenheim, *The Other World*, Cambridge: CUP, 1985, is otherwise a fascinating study.
35 M. Faraday, 'Table Turning', *Experimental Researches in Chemistry and Physics*, London: Taylor and Francis, 1859, pp. 382–91.
36 W.H. Brock, 'A British career in chemistry: Sir William Crookes, 1832–1919' in D.M. Knight and H. Kragh (eds), *The Making of the Chemist*, Cambridge: CUP, 1998, pp. 121–9.
37 W. Crookes, 'Spiritualism viewed by the light of modern science', *Quarterly Journal of Science*, 7 (1870), 316–21; W. Crookes, 'Experimental investigation of a new force', and 'Some further experiments on psychic force', *Quarterly Journal of Science*, 8 (1871), 339–49, 471–93.
38 P. Allen, *The Cambridge Apostles: the Early Years*, Cambridge: CUP, pp. 6–9.
39 A. Gauld, *The Founders of Psychical Research*, London: Routledge, 1968, tells the story very well.
40 E. Gurney, F.W.H. Myers and F. Podmore, *Phantasms of the Living*, London: SPR, 1886.
41 F. Podmore, *Studies in Psychical Research*, London: Kegan Paul, 1897.
42 F.W.H. Myers, *Human Personality and its Survival of Bodily Death*, London: Longman, 1903.
43 O. Sacks, *The Man Who Mistook his Wife for his Hat*, London: Picador, 1986, pp. 193f., 232.
44 J. Oppenheim, *The Other World*, Cambridge: CUP, 1985, p. 260.
45 F.W.H. Myers, *Human Personality and its Survival of Bodily Death* [1919], repr. Norwich: Pelegrin, 1992.

NOTES

46 F.W.H. Myers, *Human Survival and its Survival of Bodily Death*, Charlottesville, VA: Hampton Roads, 2001 (new edn of 1961 abridgement, New York: University Books).
47 C.D. Broad, *Religion, Philosophy and Psychical Research*, London: Routledge, 1953.
48 M. Knight, *Spiritualism, Reincarnation and Immortality*, London: Duckworth, 1950.
49 J.R. Watson, *The English Hymn: a Critical and Historical Study*, Oxford: OUP, 1997.
50 E.G. Sandford (ed.), *Memoirs of Archbishop Temple by Seven Friends*, London: Macmillan, 1906, esp. vol. 1, pp. 582–4, vol. 2, pp. 628–55.
51 C. Gore (ed.), *Lux Mundi: a Series of Studies in the Religion of the Incarnation*, 15th edn, London: Murray, 1899.
52 W. Thomson [Lord Kelvin], *Popular Lectures and Addresses*, London: Macmillan, 1894, vol. 2, pp. 6–131.
53 J.F.W. Herschel, *A Treatise on Astronomy*, new edn, London: Longman, 1851, p. 5.
54 D.M. Knight, *Ideas in Chemistry*, London: Athlone, 1992, chapter 12.
55 R. McCormmach, *Night Thoughts of a Classical Physicist*, Cambridge, MA: Harvard UP, 1982.
56 F. Burckhardt, D.M. Porter, J. Harvey and M. Richmond (eds), *The Correspondence of Charles Darwin*, vol. 9, Cambridge: CUP, 1994, pp. 135f.

10 HANDLING CHANCE

1 Horace, *Epistles*, bk 1, ep. 2, line 27: *Nos numerus sumus et fruges consumere nati*.
2 T.H. Huxley, 'A liberal education and where to find it' [1868] in *Science and Education by Thomas Huxley*, ed. C. Winick, New York: Citadel, 1964, pp. 72–100, esp. pp. 77–80.
3 See II Chronicles, 18: 33.
4 M.J.S. Rudwick, *Scenes from Deep Time*, Chicago: Chicago UP, 1992, pp. 30–6.
5 G. Gigerenzer, Z. Swijtink, T. Porter, L. Daston, J. Beatty and L. Krüger, *The Empire of Chance: how Probability changed Science and Everyday Life*, Cambridge: CUP, 1989; C. Sherrington, *Man on his Nature*, London: Penguin, 1955.
6 C. Sherrington, *Man on His Nature*, 2nd edn, London: Penguin, 1955.
7 W.E. Gladstone, *Studies Subsidiary to the Works of Bishop Butler*, Oxford: Clarendon Press, 1896.
8 L. Stewart, 'Radical physic in the chemical revolution', in T. Levere and G.L'E. Turner (eds), *Discussing Chemistry and Steam*, Oxford: OUP, 2002, pp. 231–45; D.A. Stansfield, *Thomas Beddoes, MD, 1760–1808*, Dordrecht: Reidel, 1984, p. 182.
9 R. FitzRoy, *The Weather Book*, London: Longman, 1863.
10 P.S. Laplace, *Essai Philosophique sur les Probabilités*, repr. Paris: Gauthier-Villars, 1921.
11 E. FitzGerald, *The Rubáiyát of Omar Khayyám* [1859], London: Folio Society, 1955, stanza 53.
12 C. Babbage, *Ninth Bridgewater Treatise*, 2nd edn, London: Murray, 1838, pp. xii–xix.
13 J. Vining, *From Newton's Sleep*, Princeton, NJ: Princeton UP, 1995.
14 W. Whewell, *Astronomy and General Physics considered with reference to Natural Theology*, Bridgewater Treatise no. 3, 1st edn, London: William Pickering, 1833, chapter 1.
15 J. Morrell and A. Thackray, *Gentlemen of Science: Early Years of the British Association for the Advancement of Science*, Oxford: Clarendon Press, 1981, pp. 291–6.
16 T.H. Lister, *First Annual Report of the Registrar-General of Births, Deaths and*

Marriages in England, London: HMSO, 1839.
17 Registrar-General, *Weekly Return Of Births and Deaths in London*, 20 (1859), no. 35, p. 1; no. 36, p. 1; no. 47, p. 1.
18 R.B. Freeman, *The Works of Charles Darwin; an Annotated Bibliographical Handlist*, 2nd edn, Folkestone: Dawson, 1977, p. 75; *Charles Darwin: a Companion*, Folkestone: Dawson, 1978, p. 220.
19 S.H. Preston and M.R. Haines, *Fatal Years: Child Mortality in late 19th-century America*, Princeton, NJ: Princeton UP, 1991.
20 T.R. Malthus, *An Essay on the Principle of Population*, London: Johnson, 1798, p. 48.
21 F. Burckhardt et al. (eds), *The Correspondence of Charles Darwin*, Cambridge: CUP, vol. 7, 1991, pp. 392, 423; vol. 9, 1994, p. 135; vol. 8, 1993, p. 81, and on *vera causa*, p. 77.
22 C. Darwin, *On the Various Contrivances by which British and Foreign Orchids are Fertilised by Insects*, London: Murray, 1862, pp. 197–8.
23 C. Lyell, *The Geological Evidences of the Antiquity of Man*, London: Murray, 1863, p. 498.
24 I. Newton, *Opticks* [4th edn, 1730], New York: Dover, 1952, p. 400.
25 J.C. Maxwell, *Scientific Papers*, ed. W.D. Niven, Cambridge: CUP, 1890, vol. 1, pp. 377–409; vol. 2, pp. 1–25, 26–78; these are mathematical, but pp. 361–77 is a more accessible lecture.
26 J.P. Joule, *Scientific Papers*, London: Physical Society, 1884, vol. 1, pp. 298–328.
27 J.C. Greene, *Science, Ideology and World View*, Berkeley, CA: Univ. of California Press, 1981, pp. 128–57.
28 J. Arbuthnot, 'An argument for Divine Providence', *Philosophical Transactions Abridged*, 5 (1809), 606–8 [1710].
29 K.E. von Baer, *De Ovi Mammalium et Hominis Genesi*, Leipzig: Vossius, 1827.
30 E. Halley, *Philosophical Transactions Abridged* [1692–3], 3 (1809), 483–91.
31 A. de Morgan, *A Budget of Paradoxes* [2nd edn, 1895], Freeport, NY: Books for Libraries, 1969; M.A. de Morgan (ed.), *Threescore Years and Ten: Reminiscences of the late Sophia Elizabeth de Morgan*, London: Richard Bentley, 1895.
32 'Bishop Howley's charge', *Quarterly Review*, 14 (1815–16), 41–2.
33 A. Kuper, 'Incest, cousin marriage, and the origin of the human sciences in 19th-century England', *Past and Present*, 174 (2002), 158–83.
34 F. Galton, *The Art of Travel, or Shifts and Contrivances available in Wild Countries* [1872], intro. D. Middleton, Newton Abbot: David and Charles, 1971.
35 J. Herschel (ed.), *A Manual of Scientific Enquiry: Prepared for the Use of Officers in Her Majesty's Navy; and Travellers in General*, 2nd edn, London: Murray, 1851 [repr., ed. D.M. Knight, Dawson, 1974].
36 F. Galton, *Hereditary Genius: an Inquiry into its Laws and Consequences*, 2nd edn, London: Macmillan, 1892. Galton notes (p. vii) that this is merely a corrected reprint of the earlier edition.
37 *Ibid.*, pp. 321, 325, 327, 152–6, 332, 337, 343.
38 *Ibid.*, p. xxvii.
39 H.F. Augstein (ed.), *Race: the Origins of an Idea, 1760–1850*, Bristol: Thoemmes, 1996.
40 F. Galton, 'Men of science: their nature and their nurture', *Proceedings of the Royal Institution*, 7 (1873–5), 227–36.
41 My copy of Galton's 1874 British Institution lecture was sent to Piazzi Smyth, Astronomer Royal for Scotland.
42 D.A. MacKenzie, *Statistics in Britain, 1865–1930: the Social Construction of Scientific Knowledge*, Edinburgh: Edinburgh UP, 1981.
43 W. Shakespeare, *The Merchant of Venice*, IV.i.88.

44 P.J. Bowler, *Reconciling Science and Religion: the Debate in Early-twentieth-century Britain*, Chicago: Chicago UP, 2001, pp. 234–5, 240, 250–51, 257, 259, 266, 273–4.
45 A. Tennyson, *In Memoriam*, ed. S. Shatto and M. Shaw, Oxford: OUP, 1982, p. 80 (section 56).
46 F. Burckhardt et al. (eds), *The Correspondence of Charles Darwin*, Cambridge: CUP, vol. 8, 1993, pp. 224, 274f, 496; vol. 9, 1994, pp. 13, 29f, 39, 51, 89, 162, 226, 238, 267, 358, 369; vol. 10, 1997, pp. 86, 117, 140; vol. 11, 1999, p. 168 (quotation).
47 C. Darwin, *On the Origin of Species*, London: Murray, 1859, pp. 60, 197, 224, 453.
48 R.L. Numbers, *Creation by Natural Law: Laplace's Nebular Hypothesis in American Thought*, Seattle: Washington UP, 1977; esp. pp. 20–36.
49 R. Chambers, *Vestiges of the Natural History of Creation* [1844], ed. J. Secord, Chicago: Chicago UP, 1994, pp. 1–28.
50 J.P. Nichol, *The Architecture of the Heavens*, London: Parker, 1850.
51 W. Huggins, *The Royal Society: or, Science in the State and the Schools*, London: Methuen, 1906.
52 D.N. Livingstone, *Darwin's Forgotten Defenders: the Encounter between Evangelical Theology and Evolutionary Thought*, Edinburgh: Scottish Academic Press, 1987; and for comparisons, R.L. Numbers and J. Stenhouse (eds), *Disseminating Darwinism: the Role of Place, Race, Religion, and Gender*, Cambridge: CUP, 1999.

11 CLERGY AND CLERISY

1 K. Gispen, *New Profession, Old Order: Engineers and German Society, 1815–1914*, Cambridge: CUP, 1989.
2 J.C. Beaglehole (ed.), *The Endeavour Journal of Joseph Banks, 1768–1771*, Sydney: Angus and Robertson, 1962; H.B. Carter, *Sir Joseph Banks, 1743–1820*, London: Natural History Museum, 1988.
3 J. Barwell-Carter and J. Hardy (eds), *Selections from the Correspondence of Dr George Johnston*, Edinburgh: David Douglas, 1892, pp. 110, 122, 135, 491, 155, 221, 243, 384.
4 T. Levere and G.L'E. Turner (eds), *Discussing Chemistry and Steam: the Minutes of a Coffee House Philosophical Society, 1780–1787*, Oxford: OUP, 2002.
5 J. Brooke and G. Cantor, *Reconstructing Nature: the Engagement of Science and Religion*, Edinburgh: T. & T. Clark, 1998, pp. 282–313.
6 J. Uglow, *The Lunar Men: the Friends who made the Future*, London: Faber and Faber, 2002.
7 C.R. Weld, *A History of the Royal Society* [1848], repr. Bristol: Thoemmes, 2000, vol. 2, pp. 103–301.
8 M.P. Crosland, *Science Under Control: the French Academy of Sciences, 1795–1914*, Cambridge: CUP, 1992.
9 J.P.R. Deleuze, *History and Description of the Royal Museum of Natural History*, Paris: Royer, 1825.
10 J.P. Poirier, *Lavoisier: Chemist, Biologist, Economist*, tr. R. Balinski, Philadelphia: Univ. of Pennsylvania Press, 1996, chapters 6, 10, 13, 17; M.P. Crosland, *Gay-Lussac: Scientist and Bourgeois*, Cambridge: CUP, 1978, chapter 8.
11 C.R. Weld, *History of the Royal Society*, vol. 2, p. 99.
12 J. Gascoigne, *Science in the Service of Empire: Joseph Banks, the British State and the Uses of Science in the Age of Revolutions*, Cambridge: CUP, 1998; R.E.R. Banks et al. (eds), *Sir Joseph Banks: a Global Perspective*, London: Kew, 1994.
13 D.G. King-Hele (ed.), *John Herschel, 1792–1871*, London: Royal Society, 1992.

NOTES

14 D.M. Knight, 'Scientific lectures: a history of performance', *Interdisciplinary Science Reviews*, 27 (2002), 217–24.
15 F.A.J.L. James (ed.), *The Correspondence of Michael Faraday*, vol. 2, London: Institution of Electrical Engineers, 1993, pp. 244ff, 286–322; on 'professional business', p. 458.
16 A. Sedgwick, *A Discourse on the Studies of the University* [1833], intro. E. Ashby and M. Anderson: Leicester: Leicester UP, 1969, p. 14.
17 H.W. Becher, 'Voluntary science in 19th-century Cambridge University to the 1850s', *BJHS*, 19 (1986), 57–87.
18 C. Wordsworth, *Annals of my Early Life, 1806–1846*, London: Longman, 1891, pp. 55–62.
19 S.T. Coleridge, *Aids to Reflection* [1825], ed. J. Beer, London: Routledge, 1993; S.T. Coleridge, *On the Constitution of the Church and State* [1829], ed. J. Colmer, London: Routledge, 1976; K. Coburn, *Inquiring Spirit: a New Presentation of Coleridge*, London: Routledge, 1951; R. Holmes, *Coleridge: Darker Reflections*, London: HarperCollins, 1998, pp. 537–59.
20 F.M. Turner, *John Henry Newman, the Challenge to Evangelical Religion*, New Haven, CT: Yale UP, 2002.
21 P. White, *Thomas Huxley: Making the 'Man of Science'*, Cambridge: CUP, 2003, pp. 1–5.
22 J. Morrell and A. Thackray, *Gentlemen of Science: Early Years of the British Association for the Advancement of Science*, Oxford: OUP, 1981; R. MacLeod and P. Collins (eds), *The Parliament of Science*, London: Science Reviews, 1981.
23 C. Babbage, *Reflections on the Decline of Science in England, and on Some of its Causes*, London: Fellowes, 1830.
24 L. Oken, *Elements of Physiophilosophy*, tr. A. Tulk, London: Ray Society, 1847.
25 P. Ackroyd, *London: the Biography*, London: Vintage, 2001, p. 683, makes this remark in the context of music halls.
26 For a close study of individual 'trajectories' in science at this time, see M.J.S. Rudwick, *The Great Devonian Controversy: the Shaping of Scientific Knowledge among Gentlemanly Specialists*, Chicago: Chicago UP, 1985, pp. 410–28.
27 J. Morrell, 'The early Yorkshire Geological and Polytechnic Society: a reconsideration', *Annals of Science*, 45 (1988), 153–67; 'Genesis and geochronology: the case of John Phillips', in C.L.E. Lewis and S.J. Knell (eds), *The Age of the Earth: from 4004BC to AD2002*, London: Geological Society, 2001. Jack Morrell is writing a biography of Phillips, to be published by Ashgate.
28 J. Phillips, *Illustrations of the Geology of Yorkshire*, 2 vols., York: Wilson, 1829/London: Murray, 1836; see W.C. Williamson, *Reminiscences of a Yorkshire Naturalist* [1896], intro. J. Watson and B.A. Thomas, Manchester: Watson and Thomas, 1985, p. 12.
29 G. Young, *Scriptural Geology: or, an Essay on the High Antiquity ascribed to the Organic Remains imbedded in Stratified Rocks*, 2nd edn, London: Simpkin Marshall, 1840.
30 D.S. Evans, T.J. Deeming, B.H. Evans and S. Goldfarb (eds), *Herschel at the Cape: Diaries and Correspondence of Sir John Herschel, 1834–38*, Austin: Univ. of Texas, 1969, p. 317.
31 A. Secord, 'Artisan botany', in N. Jardine, J.A. Secord and E.C. Spary (eds), *Cultures of Natural History*, Cambridge: CUP, 1996, pp. 378–93.
32 See M.P. Crosland, *Historical Studies in the Language of Chemistry*, 2nd edn, New York: Dover, 1978, pp. 256–81; D.M. Knight, *Ideas in Chemistry: a History of the Science*, London: Athlone, 1992, chapter 9.
33 W. Jardine (ed.), *Memoirs of Hugh Edwin Strickland*, London: Van Voorst, 1858, pp. 375–97.

34 *Ibid.*, pp. 408–17.
35 J. Coggon, 'The circular system of William Hincks', *Journal of the History of Biology* (2002), 5–42.
36 S.G. Kohlstedt, *The Formation of the American Scientific Community: the A.A.A.S. 1848–1860*, Urbana: Illinois UP, 1976.
37 R. MacLeod (ed.), *The Commonwealth of Science: ANZAAS and the Scientific Enterprise in Australasia, 1888–1988*, Oxford: OUP, 1988; R.W. Home (ed.), *Australian Science in the Making*, Cambridge: CUP, 1988.
38 P.R. Sweet, *Wilhelm von Humboldt: a Biography*, Columbus: Ohio State UP, 1980, vol. 2, pp. 3–88.
39 T.H. Broman, *The Transformation of German Academic Medicine, 1750–1820*, Cambridge: CUP, 1996.
40 J.B. Morrell, 'The chemist breeders: the research schools of Liebig and Thomas Thomson', *Ambix*, 19 (1972), 1–46; W.H. Brock, *Justus von Liebig: the Chemical Gatekeeper*, Cambridge: CUP, 1997, pp. 37–71.
41 See E. Homburg, 'Two factions, one profession: the chemical profession in German society, 1780–1870', in D.M. Knight and H. Kragh (eds), *The Making of the Chemist: the Social History of Chemistry in Europe, 1789–1914*, Cambridge: CUP, 1998, pp. 39–76.
42 D.M. Knight, 'Davy and Faraday: fathers and sons', in D. Gooding and F.A.J.L. James (eds), *Faraday Rediscovered*, Basingstoke: Macmillan, 1985, pp. 33–49; F.A.J.L. James (ed.), *The Correspondence of Michael Faraday*, vol. 1, London: Institute of Electrical Engineers, 1991, p. xxxiv; J. Hamilton, *Faraday: the Life*, London: HarperCollins, 2002, pp. 186–95.
43 P. White, *Thomas Huxley: Making the 'Man of Science'*, Cambridge: CUP, 2003, pp. 100–134.
44 C.E. McClelland, *State, Society and University in Germany, 1700–1914*, Cambridge: CUP, 1980.
45 A.J. Rocke, *Nationalizing Science: Adolphe Wurtz and the Battle for French Chemistry*, Cambridge, MA: MIT Press, 2001; G.L. Geison, *The Private Science of Louis Pasteur*, Princeton, NJ: Princeton UP, 1995.
46 E. Homburg, A.S. Travis and H.G. Schröter (eds), *The Chemical Industry in Europe, 1850–1914: Industrial Growth, Pollution and Professionalization*, Dordrecht: Kluwer, 1998; C. Meinel and H. Scholz (eds), *Die Allianz von Wissenschaft und Industrie: August Wilhelm Hofmann (1818–1892), Zeit, Werk, Wirkung*, Weinheim: VCH, 1992.
47 *Royal Commission on Scientific Instruction and the Advancement of Science*, 2 vols., London: HMSO, 1872; W.H. Brock, *Science for All: Studies in the History of Victorian Science and Education*, Aldershot: Ashgate Variorum, 1996.
48 A. Desmond, *Huxley: Evolution's High Priest*, London: Michael Joseph, 1997.
49 P. White, *Thomas Huxley: Making the 'Man of Science'*, Cambridge: CUP, 2003, pp. 62–6.
50 J.V. Jensen, 'Return to the Wilberforce–Huxley debate', *BJHS*, 21 (1988), 161–80.
51 J.W. Draper, *History of the Conflict between Religion and Science*, London: King, 1875. My copy was Herbert Spencer's, but he has merely rubber-stamped it.
52 T.S. Kuhn, 'The function of dogma in scientific research', in A.C. Crombie (ed.), *Scientific Change*, Oxford: OUP, 1963, pp. 347–69, and the discussion that follows.

12 MASTERING NATURE

1 This chapter is based on a talk I gave at Newcastle Literary and Philosophical Society on 26 September 2000, published as 'Why is science so macho?',

NOTES

Philosophical Writings, 14 (2000), 59–65.
2. C.A. Russell, *The Earth, Humanity and God*, London: UCL Press, pp. 12–18; O. Mayr, *Authority, Liberty and Automatic Machinery in Early Modern Europe*, Baltimore: Johns Hopkins UP, 1986.
3. S. Roebuck (ed.), 'Being responsible in a cosmic context', *Christ and the Cosmos*, 8 (1994) [Oxford: Westminster College].
4. G. Vancouver, *A Voyage of Discovery to the North Pacific Ocean and Round the World*, London: Robinson, 1798, vol. 2, p. 17; and cf. K. Oslund, 'Imagining Iceland: narratives of nature and history in the North Atlantic', *BJHS*, 35 (2002), 313–34.
5. See *Interdisciplinary Science Reviews*, 27 (2002), 161–247, on this theme; and also R. Hamblyn, *The Invention of Clouds*, London: Picador, 2001, pp. 4–14; M. Kwint, 'The legitimization of the circus in late Georgian England', *Past and Present*, 174 (2002), 72–115.
6. P. Bertucci and G. Pancaldi, *Electric Bodies: Episodes in the History of Medical Electricity*, Bologna: Università di Bologna, Dipartimento di Filosofia, 2001.
7. *Register of Arts and Sciences*, 1 (1824), 3–5.
8. J. Uglow, *The Lunar Men: the Friends who made the Future*, London: Faber, 2002, p. 484.
9. J.A. Paris, *The Life of Sir Humphry Davy*, London: Colburn and Bentley, 1831, p. 89, capitalised – in Davy's *Collected Works*, London: Smith Elder, 1839–40, vol. 2, pp. 323, 318–19, 314, 326 it is not.
10. P. Strathern, *Mendeleyev's Dream; the Quest for the Elements*, London: Hamish Hamilton, 2000.
11. D.M. Knight, *Humphry Davy: Science and Power*, 2nd edn, Cambridge: CUP, 1998, p. 130.
12. F.M. Turner, *John Henry Newman: the Challenge to Evangelical Religion*, New Haven, CT: Yale UP, 2002, pp. 425–36.
13. A.L. Barbauld, *The Poems*, ed. W. McCarthy and E. Kraft, Athens, GA: Georgia UP, 1994, p. 158.
14. N. Jardine, J.A. Secord and E.C. Spary (eds), *Cultures of Natural History*, Cambridge: CUP, 1996; L.N. Nyhart, *Biology Takes Form: Animal Morphology and the German Universities, 1800–1900*, Chicago: Chicago UP, 1995.
15. A. Thwaite, *Glimpses of the Wonderful: the Life of Philip Henry Gosse*, London: Faber, 2002, pp. 123, 238.
16. J.J. Thomson, *Recollections and Reflections*, London: Bell, 1936, pp. 378–9.
17. D.M. Knight, *Ideas in Chemistry: A History of the Science*, London: Athlone, 1992, pp. 157–70; M.J. Nye, *Before Big Science: the Pursuit of Modern Chemistry and Physics, 1800–1940*, New York: Twayne, 1996, pp. 147–88.
18. R.C. Lewontin, 'Genes, environment and organisms', in R.B. Silvers (ed.), *Hidden Histories of Science*, London: Granta, 1995, pp. 115–39.
19. H. Davy, *Consolations in Travel*, London: Murray, 1830, pp. 251–2, 65.
20. F. Accum, *A System of Theoretical and Practical Chemistry*, 2nd edn, London: Kearsley, 1807; and his *Chemical Reagents or Tests*, 2nd edn, London: Charles Tilt; both were advertisements as well as texts.
21. For an anecdote, in a delightfully worm's-eye-view of science, see E. Grey, *Reminiscences, Tales and Anecdotes of the Laboratories, Staff and Experimental Fields, 1872–1922*, Rothamsted Experimental Station: n.d. [1922], p. 48.
22. A.J. Rocke, *Nationalizing Science: Adolphe Wurtz and the Battle for French Chemistry*, Cambridge, MA: MIT, 2001, p. 26; D.M. Knight, *Humphry Davy*, Cambridge: CUP, 1998, p. 65.
23. M. Faraday, *Chemical Manipulation*, 3rd edn, London: Murray, 1842 [1st edn, 1827].

NOTES

24 W. Wordsworth, *Poetry and Prose*, ed. W.M. Merchant, London: Hart Davis, 1955, pp. 508–10.
25 F. Galton, 'On men of science: their nature and their nurture', *Proceedings of the Royal Institution*, 7 (1873–5), 227–36, at pp. 228–9.
26 A. Tennyson, *In Memoriam*, ed. S. Shatto and M. Shaw, Oxford: OUP, 1982, p. 115.
27 J. Sutherland, *Mrs Humphry Ward: Eminent Victorian, Pre-eminent Edwardian*, Oxford: OUP, 1990, pp. 106–31.
28 M. Beretta, *Imaging a Career in Science: the Iconography of Antoine Laurent Lavoisier*, Canton, MA: Science History, 2001.
29 F.A.J.L. James (ed.), *The Correspondence of Michael Faraday*, vol. 3, London: Institution of Electrical Engineers, 1996, pp. 164, 253–5, 264–7, 270–2, 276–7, 279–83, 291–2, 323–4; K.K. Schwarz, 'Faraday and Babbage', *Notes and Records of the Royal Society*, 56 (2002), 367–81, at pp. 369, 376–7.
30 C. Djerassi and R. Hoffmann, *Oxygen*, Weinheim: Wiley–VCH, 2001.
31 H.T. Buckle, *The Miscellaneous and Posthumous Works*, ed. G. Allen, London: Longman, 1885, vol. 1, p. 63; abstract in *Proceedings of the Royal Institution*, 2 (1854–8), 504–5.
32 P. Duhem, *The Aim and Structure of Physical Theory*, tr. P.P. Wiener, New York: Atheneum, 1962, pp. 55–103.
33 W. Babington, A. Marcet and W. Allen, *A Syllabus of a Course of Chemical Lectures read at Guy's Hospital*, London: Royal Free School, 1811; copies were interleaved for students to make notes.
34 [J. Marcet], *Conversations on Chemistry: in which the Elements of that Science are Familiarly Explained and Illustrated by Experiments*, 11th edn, London: Longman, 1828, p. v.
35 H. Davy, *Consolations in Travel*, London: Murray, 1830, p. 251.
36 [E. and S.M. Fitton], *Conversations on Botany*, 6th edn, London: Longman, 1828.
37 K.A. Neeley, *Mary Somerville: Science, Illumination, and the Female Mind*, Cambridge: CUP, 2001.
38 A. Secord, 'Botany on a plate: pleasure and the power of pictures in promoting early 19th-century scientific knowledge', *Isis*, 93 (2002), 28–57; D.M. Knight, *Zoological Illustration*, Folkestone: Dawson, 1977, and *Natural Science Books in English*, 2nd edn, London: Portman, 1989.
39 J. Lindley, *Sertum Orchidaceum: a Wreath of the Most Beautiful Orchidaceous Flowers*, London: Ridgway, 1838; J. Bateman, *The Orchidaceae of Mexico and Guatemala*, London: Ackermann, 1837–42. These are serious botanical drawings as well as being highly decorative.
40 G.C. Sauer, *John Gould, the Bird Man: a Chronology and Bibliography*, London: Sotheran, 1982; C.E. Jackson, *Bird Illustrators: some Artists in Early Lithography*, London: Witherby, 1975, pp. 39–58.
41 Preface by the editor, Richard Taylor, to *Scientific Memoirs*, 1 (1837), iii.
42 *Dictionary of Literary Biography*, New York: Gale, vol. 106, 1991, pp. 59–62.
43 H.C. Oersted, *Selected Scientific Works*, ed. and tr. K. Jelved, A.D. Jackson and O. Knudsen, intro. A.D. Wilson, Princeton, NJ: Princeton UP, 1998; *The Soul in Nature*, tr. L. and J.B. Horner, London: Bohn. 1852. There was an international conference on Oersted at Harvard in May 2002, which will be published.
44 N.L. Paxton, *George Eliot and Herbert Spencer: Feminism, Evolutionism, and the Reconstruction of Gender*, Princeton, NJ: Princeton UP, 1991.
45 Her dates were 1818–1903; see her entry in *Dictionary of National Biography*, the 1901–1910 volume.
46 See Leonard Horner's obituary, *Proceedings of the Royal Society*, 14 (1865), v–x; his

NOTES

presentation copy of *The Origin of Species* [1859] was later given by Joanna to the Natural History Museum.

47 [J.W. Croker], *Quarterly Review*, 18 (1817–18), 379, 382, 385.
48 Among recent editions are M. Shelley, *Frankenstein: or the Modern Prometheus*, ed. J.C. Oates, illus. B. Moser, Berkeley: California UP, 1994; ed. D.L. Macdonald and K. Scherf, 2nd edn, Peterborough, Ontario: Broadview, 1999, which reprints original reviews (some more perceptive than Croker's).
49 M. Seymour, *Mary Shelley*, London: Murray, 2000, pp. 153–64.
50 H.L. Malchow, 'Frankenstein's monster and images of race in 19th-century Britain', *Past and Present*, 139 (1993), 90–130.
51 T.H. Broman, *The Transformation of German Academic Medicine, 1750–1820*, Cambridge: CUP, 1996, p. 165.
52 *Register of Arts and Sciences*, 1 (1824), 3–6; C. Blondel, 'Animal electricity in Paris' in M. Bresadola and G. Pancaldi (eds), *Luigi Galvani International Workshop*, Bologna: Bologna University, 1999, pp. 201–2.
53 T.H. Levere, *Science and the Canadian Arctic: a Century of Exploration, 1818–1918*, Cambridge: CUP, 1993.
54 J. Cook, *A Voyage Towards the South Pole, and Round the World*, London: Strahan and Cadell, 1777, vol. 1, p. 268; note for 30 January 1774.
55 H. Collins and T. Pinch, *The Golem: What you should know about Science*, and *The Golem at Large: What you should know about Technology*, Cambridge: CUP, 1993 and 1998.
56 O. Moscucci, *The Science of Woman: Gynaecology and Gender in England, 1800–1929*, Cambridge: CUP, 1990.
57 *História Ciências Saúde Manguinhos*, 8 supplemento (2002), 809–1135.
58 L.R. Hiatt and R. Jones, 'Aboriginal conceptions of the workings of nature', in R. Home (ed.), *Australian Science in the Making*, Cambridge: CUP, 1988.
59 T.R. Dunlap, *Nature and the English Diaspora: Environment and History in the United States, Canada, Australia and New Zealand*, Cambridge: CUP, 1999.
60 R. Brown, 'General remarks, geographical and systematical, on the botany of Terra Australis', in M. Flinders, *A Voyage to Terra Australis*, London: Nicol, 1814, vol. 2, pp. 533–613.
61 J. Browne, 'Biography and empire', in N. Jardine, J.A. Secord and E.C. Spary (eds), *Cultures of Natural History*, Cambridge: CUP, 1996, pp. 305–21; and, focused on Brazil, the special issue of *História Ciênces Saúde Manguinhos*, 8 supplemento (2001), 809–1135.
62 J.B. Boussingault, *Rural Economy, in its Relations with Chemistry, Physics and Meteorology: or an Application of the Principles of Chemistry and Physiology to the Details of Practical Farming*, tr. G. Law, 2nd edn, London: Bailliere, 1845.
63 W.H. Brock, *Justus von Liebig: the Chemical Gatekeeper*, Cambridge: CUP, 1997, pp. 145–82.
64 A.D. Hall, *The Book of the Rothamsted Experiments*, ed. E.J. Russell, 2nd edn, London: Murray, 1917; E. Grey, *Reminiscences, Tales and Anecdotes of the Laboratories, Staff and Experimental Fields, 1872–1922*, Rothamsted Experimental Station, n.d. [1922].
65 J.H. Brooke and G. Cantor, *Reconstructing Nature: the Engagement of Science and Religion*, Edinburgh: T. and T. Clark, 1998, pp. 314–46.
66 E. Homburg, A.S. Travis and H.G. Schröter, *The Chemical Industry in Europe, 1850–1914: Industrial Growth, Pollution, and Professionalization*; and A.S. Travis, H.G. Schröter, E. Homburg and P.J.T. Morris, *Determinants in the Evolution of the European Chemical Industry, 1900–1939*, Dordrecht: Kluwer, 1998.
67 J. Krige and D. Pestre (eds), *Science in the Twentieth Century*, Amsterdam: Harwood, 1997.

68 J. Agar, *Science and Spectacle: the Work of Jodrell Bank in Post-War British Culture*, Amsterdam: Harwood, 1998.
69 L. Guzzetti (ed.), *Science and Power: the Historical Foundations of Research Policy in Europe*, Luxembourg: EU, 2000.
70 L. Jordanova, *Sexual Visions: Images of Gender in Science and Medicine between the 18th and 20th Centuries*, Hemel Hempstead: Harvester Wheatsheaf, 1989.
71 E. Shils and C. Blacker, *Cambridge Women: Twelve Portraits*, Cambridge: CUP, 1996; S.G. Kohlstedt (ed.), *History of Women in the Sciences: Readings from Isis*, Chicago: Chicago UP, 1999.

13 MEANING AND PURPOSE?

1 H. de Almeida, *Romantic Medicine and John Keats*, Oxford: OUP, 1991.
2 B.R. Haydon, *Autobiography and Journals, 1786–1846*, ed. M. Elwin, London: Macdonald, 1950, p. 317.
3 J.-M. Lévy-Leblond, 'Two cultures or none?', *Pantaneto Forum*, 8 October 2002; and my 'Working in the glare of two cultures', *Interdisciplinary Science Reviews*, 23 (1998), 156–60.
4 P. White, *Thomas Huxley: Making the 'Man of Science'*, Cambridge: CUP, 2003, pp. 67–99.
5 R. Fox and G. Weisz (eds), *The Organization of Science and Technology in France, 1808–1914*, Cambridge: CUP, 1980.
6 R.W. Home, 'The Royal Society and Empire: the colonial and Commonwealth Fellowship, part 1, 1731–1847', *Notes and Records of the Royal Society*, 56 (2002), 307–32.
7 BAAS, *Narrative and Itinerary of the Australian Meeting, 1914*, London: BAAS, 1915.
8 C. Meinel and H. Scholz, *Die Allianz von Wissenschaft und Industrie: August Wilhelm Hofmann (1818–92), Zeit, Werk, Wirkung*, Weinheim: VCH, 1992.
9 W.H. Brock, *Science for All: Studies in the History of Victorian Science and Education*, Aldershot: Ashgate Variorum, 1996.
10 S.E. Koss, *Sir John Brunner: Radical Plutocrat, 1842–1919*, Cambridge: CUP, 1970; W. von Siemens, *Inventor and Entrepreneur: Personal Recollections*, London: Lund Humphries, 1966.
11 P.W. Hammond and H. Egan, *Weighed in the Balance: a History of the Laboratory of the Government Chemist*, London: HMSO, 1992.
12 E. Crawford, *Nationalism and Internationalism in Science, 1880–1913: Four Studies of the Nobel Population*, Cambridge: CUP, 1992.
13 B. Stewart and P.G. Tait, *The Unseen Universe: or Physical Speculations on a Future State*, 6th edn, London: Macmillan, 1876, p. 25.
14 D. Hartley, *Observations on Man, his Frame, his Duty, and his Expectations*, London: Richardson, 1749, vol. 2, p. 245.
15 J.A. Paris, *The Life of Sir Humphry Davy*, London: Colburn and Bentley, 1831, p. 263.
16 H. Hellman, *Great Feuds in Science: Ten of the Liveliest Disputes Ever*, New York: Wiley, 1998.
17 F. Franks, *Polywater*, Cambridge, MA: MIT Press, 1981; A. Kohn, *False Prophets*, Oxford: Blackwell, 1986.
18 C.M.C. Haines (ed.), *International Women in Science: a Biographical Dictionary to 1950*, Oxford: ABC–CLIO, 2002.
19 Togo Tsukahara, *Affinity and Shinwa Ryoku: Introduction of Western Chemical Concepts in Early-19th-century Japan*, Amsterdam: Gieben, 1993.

NOTES

20 M. Watanabe, *Science and Cultural Exchange in Modern History: Japan and the West*, Tokyo: Hokusen-Sha, 1997, esp. pp. 181–93.
21 R. Pineau (ed.), *The Japan Expedition 1852–4: The Personal Journal of Commodore Matthew C. Perry*, intro. S. Morrison, Washington, DC: Smithsonian, 1968.
22 M.H. Price, *Mathematics for the Multitude? A History of the Mathematical Association*, Leicester: Mathematical Association, 1994; A. Lundgren and B. Bensaude-Vincent (eds), *Communicating Chemistry: Textbooks and their Audiences*, Canton, MA: Science History Publications, 2000.
23 R.L. Numbers and J. Stenhouse (eds), *Disseminating Darwinism: the Role of Place, Race, Religion, and Gender*, Cambridge: CUP, 1999.
24 A. Koestler, *The Case of the Midwife Toad*, London: Hutchinson, 1971.
25 P.J. Bowler, *Reconciling Science and Religion: the Debate in early-20th-century Britain*, Chicago: Chicago UP, 2001.
26 A. Thwaite, *Glimpses of the Wonderful: the Life of Philip Henry Gosse*, London: Faber, 2002, p. 139.
27 F.J. Tipler, *The Physics of Immortality*, New York: Doubleday, 1994.
28 W.H. Brock, 'A British career in chemistry: Sir William Crookes (1832–1919)', in D. Knight and H. Kragh (eds), *The Making of the Chemist: the Social History of Chemistry in Europe, 1789–1914*, Cambridge: CUP, 1998.
29 M. Perutz, *I Wish I'd Made You Angry Earlier: Essays on Science and Scientists*, Oxford: OUP, 1998.
30 I. Hargittai, *Candid Science: Conversations with Famous Chemists*, London: Imperial College, 2000.
31 J. Krige and D. Pestre (eds), *Science in the Twentieth Century*, Amsterdam: Harwood, 1997.
32 P.J.T. Morris (ed.), *From Classical to Modern Chemistry: the Instrumental Revolution*, London: Royal Society of Chemistry, 2002.
33 M. Faraday, *Chemical Manipulation* [1842], London: Routledge, 1998, as volume 5 of the series of facsimiles *The Development of Chemistry, 1789–1914*, ed. D.M. Knight.
34 O. Sacks, *Uncle Tungsten: Memories of a Chemical Boyhood*, London: Picador, 2001.
35 P. Levi, *The Periodic Table*, tr. R. Rosenthal, London: Michael Joseph, 1985.
36 D.M. Knight, 'Theory, practice and status: Humphry Davy and Thomas Thomson' in G. Emptoz and P. Achevez (eds), *Between the Natural and the Artificial: Dyestuffs and Medicines*, Turnhout: Brepols, 2000, pp. 49–58.
37 C. Reinhart, *Chemical Sciences in the 20th Century: Bridging Boundaries*, Weinheim: Wiley, 2001.
38 E. Keinan and I. Schechter (eds), *Chemistry for the 21st century*, Weinheim: Wiley–VCH, 2001.
39 P. Harman and S. Mitton (eds), *Cambridge Scientific Minds*, Cambridge: CUP, 2002.
40 S.T. Coleridge, *Selected Poems*, ed. R. Holmes, London: Harper Collins, 1996, p. 180.
41 J. Tyndall, *Fragments of Science*, 10th impr., London, 1899, vol. 2, pp. 1–7, 40–5.
42 I. Newton, *Opticks* [1730], New York: Dover, 1952, p. 369.
43 K. Barth, *Dogmatics in Outline*, tr. G.T. Thomson, London: SCM, 1949, pp. 50–8.
44 D.S. Wilson, *Darwin's Cathedral: Evolution, Religion, and the Nature of Society*, Chicago: Chicago UP, 2002.
45 G. Cantor, *Michael Faraday: Sandemanian and Scientist*, London: Macmillan, 1991.
46 A. Thwaite, *Glimpses of the Wonderful: the Life of Philip Henry Gosse*, London: Faber, 2002.
47 J. Browne, *Charles Darwin: Voyaging*, London: Cape, 1995, p. 415.

NOTES

48 T.H. Huxley, 'The coming of age of *The Origin of Species*' in *Science and Culture, and Other Essays*, London: Macmillan, 1881.
49 R. Williams, 'Why I believe in the Bible', *Church Times*, 18 October 2002, p. 10.
50 R.G.A. Dolby, *Uncertain Knowledge: an Image of Science for a Changing World*, Cambridge: CUP, 1996.
51 S. Gilley, *Newman and his Age*, London: Darton, Longman & Todd, 1990, pp. 223–37; F.M. Turner, *John Henry Newman: the Challenge to Evangelical Religion*, New Haven, CT: Yale UP, 2002, pp. 561–86.
52 D. Brown, *Tradition and Imagination: Revelation and Change*, Oxford: OUP, 1999.
53 P. Corsi, *Science and Religion: Baden Powell and the Anglican Debate, 1800–1860*, Cambridge: CUP, 1988, pp. 209–28.
54 What follows is largely taken from a lay sermon delivered at Michaelmas 2002 in St Oswald's Church, Durham.
55 S. Prickett, *Origins of Narrative: the Romantic Appropriation of the Bible*, Cambridge: CUP, 1996, pp. 216–20.
56 Genesis, 28: 11–15.
57 Numbers, 22: 21–35.
58 Revelation, 2: 1.
59 Genesis, 32: 24–30.
60 Revelation, 12: 7–9.
61 S.T. Coleridge, *Aids to Reflection*, ed. J. Beer, London: Routledge, 1993, p. 107.
62 A. Thwaite, *Glimpses of the Wonderful: the Life of Philip Henry Gosse*, London: Faber, 2002, p. 224.
63 P.J. FitzPatrick, *In Breaking of Bread: the Eucharist and Ritual*, Cambridge: CUP, 1993.

Index

Abernethy, John 121
academicians 154
Academies of Sciences 13, 17, 38, 58, 154
actualism 56
Adams, Douglas 183
Agassiz, Louis 70
aggression 113–14
agnosticism 6, 86–91, 111, 124–5,127, 131, 185, 196 *see also* naturalism
Aldrin, Buzz 2
Ali, Hyder 145
allegory 4
Alps 65, 70, 89–91
altruism 107
Ampère, André-Marie 52
anatomy 44
anatomy theatres 169
Anderson, Sergeant 172
angels 194–5
Anglicans 3–4, 27–34, 39, 42–52, 57–8, 61–73, 74–82, 123, 131, 151–7, 164–6
animals 45–7, 81–2, 182
Annual Report of the Registrar-General 141
anthropic principle 187
anthropocentric teleology 23
apparatus 172
apprenticeship 152–3, 163
Arbuthnot, John 143–4, 187
Aristotle 20, 22, 26, 78, 138, 140
Armstrong, Neil 2
Arnold, Matthew 51
aspiration 2
astrology 24
Astronomer Royal 38
astronomy 21, 22–3, 24, 33, 46, 49, 187; nebular hypothesis 149–50
atheism 8–9, 15, 22, 51–2, 85 *see also* agnosticism, naturalism
Athenæum Club 155
atomic symbols 161
atoms 11–12, 20, 129
Australia 16, 180
awe 86, 186–8

Babbage, Charles 47, 49–51, 101, 158, 193

Bacon, Francis 17, 46, 58, 97, 123, 167; Bible and Nature 23, 24, 79
Baden-Powell, R. 73, 194
Baer, Carl von 102
Balfour, Arthur J. 125–9, 192, 193
Banks, Joseph 16–19, 108, 153, 180, 187; Royal Society 16, 17, 18, 41–2, 59, 155
Barbauld, Laetitia 170
barnacles 104–5
Barnes, E.W. 148
Barrington, Shute 27, 63
Barth, Karl 107, 191
Bateman, James 177
Bazalgette, Joseph 120
Becquerel, Henri 128
Beddoes, Thomas 13–14, 76, 138–9
belief *see* faith
Bell, Charles 44
bell curve 103, 148
Bentham, Jeremy 12
Bentley, Richard 21–2
Berthollet, C.L. 113
Berzelius, Jacob 161
Bible 3–4, 14, 93, 192; Genesis 68, 71, 72, 167–8; geology and 60–1; Kirby 46–7, 81
Bible Belt 6
biblical criticism 4
biology 98
Blake, William 3–4, 195
blasphemy 121
Blomfield, Charles 43
Bohn, Henry 40, 77, 177
Bohr, Niels 136
Boltzmann, Ludwig 143
books 76–7; illustrations 56, 77–8; illustrators 177; prices 40
Boscovich, Roger 11–12
botany 94–7
Boulton, Matthew 12, 154
Bowler, Peter 148, 186–7
Boyle, Robert 21, 22, 30, 37, 82
Brewster, David 159
Bridgewater, Earl of 42–3
Bridgewater Treatises 42–51, 72; authors 43–5; Babbage 49–51; Buckland 67–70;

INDEX

Kirby 45–7, 80–1; Whewell 47–9
British Association for the Advancement of Science (BAAS) 72–3, 141, 158, 158–63, 164–6, 184
British Museum 163
Broad, C.D. 134
Brown, David 193
Brown, Robert 180
Browning, Elizabeth Barrett 130
Buckland, William 44, 52, 54, 55–7, 62, 67, 120, 153, 156, 192, 193; and Agassiz's work on glaciers 70; BAAS 73; Bridgewater Treatise 67–70; direction in history 93; Kirkdale Cavern discoveries 55–7, 159; Noah's Flood 53, 55, 56, 63–4, 72
Buckle, Henry 174–5
Burke, Edmund 11, 19
Butler, Joseph 25, 138
Byron, Ada 174
Byron, Lord 178

Cabinet Cyclopedia, The 77, 78–80, 81
Calvinism 9, 61
Cambridge Platonists 22
Cambridge University 47, 58–9, 156–7
Canning, George 97
Carlyle, Thomas 88
Carpenter, W.B. 102, 131
Census of 1851 134
Chalmers, Thomas 43, 44, 61, 121
Chambers, Robert 100–4
Chambers, William 100
Chambers' Edinburgh Journal 100–1
chance 137–50, 191
change without evolution 66–7
chemist-breeding 162–3
chemistry 12, 19, 44, 105, 136, 171, 189; apparatus 172; as entertainment 169
Chesterton, G.K. 127
Chevalier, Temple 156
Children, J.G. 78
cholera 119–21
chronometers 30, 82
Clapham Sect 61
Clark, William 54, 187
Clausius, Rudolf 143
cleanliness 119–22
clergy 39, 151–2; education of 156–7
clerisy 150, 151–66
clockmaker, God as 21, 30, 82
coal 69
Cockburn, William 63, 159
coffee-house philosophical society 153–4
Coleridge, Samuel Taylor 5, 47–8, 82, 83, 85, 190, 196; and the clerisy 157–8; Lay Sermons 109–10; Pantisocracy 14
common sense 117–19
computer 50–1
Comte, Auguste 52, 111, 119, 125
Constable, John 44
contrivances 30–4, 82
Cook, James 16, 30, 179, 187

Cooper, Thomas 63
coprolites 69
creation 60–1, 150, 191; geology and 62–72
creeds 10,
Crookes, William 130–1, 131–2, 155, 171, 188, 189, 193
Cuddesdon theological college 157
Curie, Marie 128
Cuthbert, St 151
Cuvier, Georges 42, 54, 56, 70, 98, 99, 138

d'Alembert, Jean 8
Dalton, John 40, 73, 155, 158, 161
D'Arcy, Charles 148
Darwin, Charles 1, 5, 29, 80, 106, 108, 117, 136, 173, 187, 193; barnacles and degeneration 104–5; chance and natural law 140, 141–2, 143, 149; *Descent of Man* 107; funeral 53; Huxley and 53, 71–2, 114–15; and Lyell 53, 65; natural selection 53, 114–15, 134, 141–2; ordinand 155, 157; *Origin of Species* 96, 99, 114–15; refutation of Bell 44; and *Vestiges* 101
Darwin, Erasmus 12, 34, 94–7, 106, 154
Darwinism 6; social 6, 96, 125
Daubeny, Charles 73
David, Jacques Louis 174
Davy, Humphry 2, 16, 17, 62, 63, 78, 88, 153, 155, 178, 189; aggression 113–14; awesome science 82–6; Coleridge and 109–10; *Consolations in Travel* 65–6, 84–5; dualism 193; and Faraday 86–7, 163, 172, 185; miners' safety lamp 41, 59, 86, 180; pantheism 86–8, 90–1; and Peel 73, 155; physical fitness 172; Priestley and 13–14; prize from Academy of Sciences 59; Royal Institution lectures 76, 84, 112, 169–70; Royal Society 42, 43, 56–7, 59
Dawkins, Richard 192
death statistics 141
deduction 31, 48, 174–5
degeneration 104–5
Deism 15
Demiurge 22
Democritus 20, 138
demon 143
Derham, William 23–4
Descartes, René 26, 37
Design 20–36
Diderot, Denis 8
digestion 44
diluvian gravel layer 64, 66–7
dinosaurs 55
direction, history and 92–107
disease 119–22
Dissenters 9, 39, 57–8, 61–2, 75–6, 154, 156, 158; and education 76
Dissenting Academy, Hackney 13
divine providence 143–5
doctrines 5
Douai Roman Catholic seminary 9
doubt 124, 125–6

226

INDEX

Drake, Miss 177
Draper, John William 18, 116, 165
Drummond, Henry 107
Duhem, Pierre 175
Durham, University of 157

Earth, age of 105
Ecclesiastical Commissioners 75
eco-theology 167–8
Eddington, Arthur 7, 187
Edinburgh University 59
education 74–5, 76, 164; clergy 156–7; girls and science 175–6; higher 41; March of Mind 40–1, 76–7; public lecturing 39–40, 107, 108–22
Einstein, Albert 128
electricity 10–11
embryonic development 102
energy, conservation of 105
Enlightenment 15, 16
entertainment, science as 169–70
enthusiasm 190–1
Epicurus 20, 138
epidemic disease 119–22
error curve 103
Essays and Reviews 73, 74, 194
establishment 75–6, 158
ethical societies 119
ethics in research 181–2
Euclid 31
Euclidean geometry 136
eugenics 146–7, 148
Euler, Leonhard 23
evangelicals 3–4, 61–4, 75
Evans, Mary Ann (George Eliot) 111, 177
evolution 67, 92–107, 117; Wilberforce-Huxley debate 164–6
exotic plants and animals 180–1
experiment 170–1, 172–3
external nature 43–4
extinct species 54–5, 55–7, 68–9

faith 5; knowledge and 123–36; leap of 34–5, 47; making sense of belief 192–4
falsifiability criterion 142
Faraday, Michael 73, 90, 108, 113, 123, 135–6, 153, 155, 158, 174, 189, 192; and Davy 86–7, 163, 172, 185; psychical research 130; Royal Institution discourses 114
Farr, William 141
fecundity 33
fertilisers 181
Festival of the Supreme Being 58
Fichte, J.G. 34
First Cause 14, 188
Fisher, Ronald A. 148, 181
Fitton, Elizabeth 176
Fitton, Sarah 176
FitzRoy, Robert 44, 114, 139
Flanders, Michael 7
Flinders, Matthew 180

flood, biblical 53, 54, 63–4, 67–8, 72
Fontenelle, Bernard le B. de 38
foreign species 180–1
Foster, Michael 118
Fownes, George 102
Fox, William Darwin 155
fragmentation of science 18, 41–2
France 52, 58, 113, 154; Academy of Sciences 13, 17, 38, 58; Revolution 4, 7, 8–9, 12–13, 15, 84, 183
Franco-Prussian War 88
Frankenstein (Shelley) 87–8, 169, 178–80
Franklin, Benjamin 27
Franklin, John 179
fraternity 84
Frere, Hookham 97
Froude, Hurrell 158
functional teleology 23
fundamentalism 6, 190, 196

Galen 138
Galilei, Galileo 20, 37, 60
Galton, Francis 145–8, 151, 173, 191
Galvani, Luigi 169
gases 10, 142, 143
Gauss, K.F. 103
Gay-Lussac, J.J. 86, 113
Genesis 68, 71, 72, 167–8
genetics 186–8
genius, hereditary 145–7
Geological Society 18, 41–2, 59
geology 19, 52, 53–73, 172–3
geometrical theorems 31
Germany 60, 158–9, 184, 188
Gibbon, Edward 15, 19, 35
Gilbert, Davies 43
Gisborne, Thomas 3–4
glaciers 70
Gladstone, William Ewart 126
Glaisher, James 88
Glasgow University 59
Gombrich, E.H. 3
Gore, Charles 135
Gosse, Phillip Henry 67, 108, 187, 192, 196
Gould, Elizabeth 177
Gould, John 177
government-sponsored institutions 184
gravity 11
Gray, Asa 149, 193
Great Chain of Being 79–80
Great Stink 120
'greatest happiness of the greatest number' 12
Gregory, William 130
Gurney, Edmund 132, 133

Haber, Fritz 181
Hackney Phalanx 13
Hallam, Arthur 104
Halley, Edmond 144
Hamilton, William 19, 64
hands 44
Harcourt, Vernon 72, 159

INDEX

Hare, Maurice 92–3
Harrison, John 30, 82
Hartley, David 185
Hartley, L.P. 92
heat 142–3
Helvetius, Claude-Adrien 8–9
hereditary genius 145–7
Herschel, Caroline 174
Herschel, John 42, 47, 78, 104, 145, 160, 165; Darwin's *Origin of Species* 136, 142; man as interpreter of nature 123–4; Royal Society presidential candidate 50, 155
Herschel, William 174
Hertz, Heinrich Rudolf 128
high-church science 74–91
higher pantheism 82–3, 86–8
Hillary, Edmund 2
history 92–107
Hitchcock, Edward 63
Holbach, Paul (Baron) 8–9
Home, D.D. 130, 131
Hooker, Joseph 108, 111, 165
Horner sisters 178
Howard, Luke 44
Howley, William 43
Huggins, William 150
Humboldt, Alexander von 42, 177–8, 187
Hume, David 15, 19, 25–6, 32, 50
Hutton, James 54, 64
Huxley, Aldous 134
Huxley, Thomas Henry 26, 80, 86, 104, 106, 108, 110–12, 137, 151, 183, 192, 193; and Darwin 53, 71–2, 114–15; Darwinism 6; debate with Wilberforce 164–6; disease 121–2; lecturing 110–11, 114–19, 163; positivism 52, 111–12; and Tennyson 83; Thomson's calculations 105
hyenas 55, 56
hypnosis 133

Ice Age 70–1
illustrators 177
immortality 10, 129–30, 188
induction 174–5
industrial laboratories 181, 188
informed consent 181–2
Inge, W.R. 148
institutes 163–4
instrumentation 189–90
insurance 144–5
invertebrates 97–8
Islam 29

Japan 185–6
Jeans, James 7, 187
Jenyns, Leonard 155
Jodrell Bank radio-telescope 181
John, gospel of 4
John of the Cross, St 2
Johnson, Samuel 22
Johnston, George 153
Johnston, J.F.W. 159

Jones, William (of Nayland) 45, 76, 81
Joule, James 143, 160
joy 86, 187
judgment 119–20

Kaiser Wilhelm Institutes 184
Kammerer, Paul 186
Kant, Immanuel 14, 26, 34, 47
Kay, James 121
Keats, John 183
Keble, John 47, 62, 158
Kekulé von Stradonitz, Friedrich 170
Kelvin, Lord (William Thomson) 101, 105, 114, 127–8, 135, 160, 171
Kepler, Johannes 21
Kew Gardens 180
Kidd, John 43–4
King's College, London 67, 71
Kirby, William 44, 45, 45–7, 76, 80–2, 192
Kirkdale Cavern 55–7, 159
knowledge 5; and faith 123–36
Kuhn, Thomas 6

Laboratory of the Government Chemist 184
Lamarck, Jean Baptiste 46, 54–5, 81, 97–100
Lancet, The 102
Laplace, Pierre Simon 46, 47, 48, 81, 103, 139–40, 149
Lardner, Dionysius 77, 78
laughing gas 13
Lavoisier, Antoine 12, 13, 17, 41, 113, 143, 153
Lavoisier, Mme 174
Law, Edmund 28
Lawrence, William 121
laws 191–2; chance and 49, 137–50
lay preachers 108–9
Leach, William 77, 78
leap of faith 34–5, 47
lecturing, public 39–40, 107, 108–22
Leibniz, Gottfried Wilhelm 11
Lessing, G.E. 34
Leucippus 20
Levi, Primo 189
Lewis, Meriwether 54, 187
liberal education 156–7
Liebig, Justus 67, 162, 181
Lindley, John 177
Linnaeus, Charles 3
Linnean Society 41, 45, 78
Lister, T.H. 141
Literary and Philosophical Societies 40, 42, 76
lithography 56, 77–8
Locke, John 20
Lodge, Oliver 7, 127, 134
London University 41, 71
London Zoo 180
Longman, Thomas 77
Louis XIV, King 21
love 85, 88, 123,124
Loyola, Ignatius 2
Lucretius 20, 94, 138

228

INDEX

Lunar Society of Birmingham 12, 40, 154
Lyell, Charles 63, 67, 69–70, 93, 105, 114, 149; geology 64–5; influence on Darwin 53, 65; Lamarck 99; prediction of fossil evidence 142
Lyttleton, May 126

macho nature of science 166, 167–82
McCausland, Dominick 63
MacKenzie, Donald 148
Maillet, Benoit de 97
Malthus, Thomas 141
Manchester 13
Mansel, Henry 126
Mantell, Gideon 55
Marcet, Jane 175–6
March of Mind 40–1, 76–7
Martineau, Harriet 177
Martineau, James 127
mastering nature 167–82
materialism 2, 8–19, 85; Priestley and 9–11
mathematics 47, 48, 50, 135
Mather, Cotton 23
matter, theories of 20, 128–9; *see also* atoms
Matthews, W.R. 148
Maxwell, James Clerk 136, 142, 143
meaning, and purpose 183–96
Mechanics' Institutes 40, 77
mechanisms 30–4, 82
medical ethics 181–2
medicine 38
mediums 130, 131
Mendeleyev, Dmitri Ivanovich 170
Merino sheep 16, 180
Metaphysical Society 127
meteorology 44, 139
Methodism 14, 18, 34, 61
miasma 120
Michelson-Morley experiment 128
militant secularism 51–2
military-industrial complex 181, 184
Milton, John 61
mimosa 94–5
miners' safety lamp 41, 59, 86, 180
miracles 5, 194
Moivre, Abraham de 139
morality 28
More, Hannah 76
Morgan, Augustus de 100, 144–5
Moses, Stainton 130
mountaineering 89–90
multiple personalities 134
Murchison, Roderick Impey 67, 70
museums 163–4
Myers, Frederic 132, 133–4
mysticism 2

Napoleon 9, 113, 114, 140, 175
NASA space programme 181
Natural History Museum 163
natural selection 53, 107, 114–15, 134, 141–2
natural theology 1, 6, 19, 20–36, 107, 192;

Gisborne 3–4; Paley 19, 27–34, 79
naturalism, scientific 125, 127, 185–6
nebular hypothesis 149–50
Neeley, Kathryn 176
Newman, John Henry 47, 62, 93, 158, 193, 196; conversion to Roman Catholicism 71, 124
Newton, Isaac 11, 20, 26, 142, 167, 175, 191, 192; *Opticks* 37; *Principia* 21–2, 37
Nichol, J.P. 101, 150
Nietzsche, F. 111
Nightingale, Florence 121
Noah's Flood 53, 54, 63–4, 67–8, 72
normal distribution 103–4
Numbers, Ron 150

observation 170–1
Oersted, Hans Christian 177, 178
Oken, Lorenz 159
order 21–2
ordination 151–3, 162
Otté, Elise 177–8
outmoded ideas 193–4
Owen, Richard 55, 114, 165
Oxford Movement (Tractarians) 5, 47, 61–2
Oxford museum 163
Oxford University 58–9, 71, 157
oxygen 13
Oxygen (Djerassi and Hoffmann) 174

Paley, William 3, 5, 19, 26, 27–34, 35–6, 58–9, 82, 191; *Natural Theology* 27, 29, 30–4, 42, 58, 79, 95
pangenesis 146
pantheism 52, 82–3, 86–91
Pantisocracy 14
Parkes, Samuel 76
Parkinson, James 54
Pasteur, Louis 164
patronage 17
Pattison, Mark 74
Paul, St 28–9, 31
Pauli, Wolfgang 171
Peacock, George 47
Pearson, Karl 148
Peel, Robert 73, 155
Pepys, Samuel 30
phantasms of the living 132
PhD degree 162–3
Phillips, John 159
philosophes 4, 8
phlogiston theory 13
physics 47–9, 135–6, 171
Pickering, William 43
piety 51–2
Planck, Max 128
planets 21, 22
Plato 20, 138
Playfair, John 54, 64
Pluche, N.A. 23
Podmore, Frank 132, 133
poetry 94–7

229

INDEX

point-atom theory 11–12
polar exploration 179
Polkinghorne, John 192
polytechnics 184
Pope, Alexander 94
Popper, Karl 142
popularisers 39–40, 108–122, 190
portable laboratories 176
positivism 52, 111, 119, 125
power 167
prayer 15,191
predators 32
prediction 142
Priestley, Joseph 2, 7, 9–14, 19, 154, 192; and Banks 16, 17–18, 19; Birmingham 12–13; exile to USA 13, 58; and materialism 9–11
probability 25, 137–50
professions 151–3
programmable computer 50–1
Prout, William 44
psychical research 130–4
Ptolemy 21
public lecturing 39–40, 107, 108–22
purpose: chance and 137–8; meaning and 183–96
Pusey, E.B. 60, 62, 68, 73, 158
Pythagoras' theorem 31

quantum theory 136
Quetelet, Lambert 101, 103–4, 148
quinary system 79–80, 161–2

radicalism 14
Raffles, Stamford 180
random genetic mutation 107
Raven, Charles 7, 107
Ray, John 22
Ray Society 153
Rayleigh, Lord 127, 128, 129
Rayleigh, Lord (second Lord) 129
reason 24–5
Reform Act 1832 62, 158
Registrar-General 141
Reichenbach, Karl von 130
Reign of Terror 8, 13
Religion of Humanity 52
research 38–9
research students 164
research universities 184
resurrection of the body 10
reverie 170
ritual 72–3
Robson, E.R. 121
Rochester, Bishop of 75
Roget, Peter Mark 44
Romantic Movement 15, 37, 82–3
Röntgen, Wilhelm Konrad von 128
Ross, Thomasina 177
Rosse, Lord 150
Rothamsted experimental station 181
rotten boroughs 62
Rousseau, Jean-Jacques 8

Royal Commission on Scientific Instruction and the Advancement of Science 164
Royal Institution 40, 42, 59, 114
Royal Society 13, 40, 154–6, 184; Banks's presidency 16, 17, 18, 41–2, 59, 155; Bridgewater Treatises 42–3; Copley Medal 56–7
Rozier, Pilatre de 95
Ruskin, John 52, 57, 114
Russell, Bertrand 129
Russell, William 196

Sabine, Edward 108, 178
Sabine, Elizabeth 178
Sacks, Oliver 189
safety lamp, miners' 41, 59, 86, 180
Saint-Hilaire, Etienne Geoffroy 99
Salisbury, Lord 125, 126
Sandemanian sect 86
sanitary engineering 120–1
scepticism 15–16, 125–6; Hume 25–6
Schleiermacher, Friedrich 34–6
Schuster, Arthur 128
scientific naturalism 125, 127, 185–6
scientism 7
scientists 147–8, 158
Scott, David 150
Scott, Robert Falcon 179
Scott, Walter 82, 92
séances 130, 131
Secord, Jim 100
secularism, militant 51–2, *see also* naturalism
Sedgwick, Adam 64, 80, 104, 156, 172–3
sermonising, lay 107, 108–22
Shelburne, Lord 12
Shelley, Mary, *Frankenstein* 87–8, 169, 178–80
Shelley, Percy 178
Sherrington, Charles 138
Sidgwick, Henry 125–6, 130, 131, 133, 193
Sidgwick, Nora 134
Singh, Runjeet 145
Smith, William 59, 159
snakes 23
Snow, C.P. 7
Snow, John 120
Social Darwinism 6, 96, 125
societies 39–41
Society for Psychical Research (SPR) 131–3, 134
Society for the Diffusion of Useful Knowledge 76
Society for the Promotion of Christian Knowledge 76
Solander, Daniel 16
solar system 21, 22, 46
Somerville, Mary 2, 5, 42, 176, 177
Southey, Robert 14
specialised societies 18, 41–2
spectroscopy 189
Spence, William 45
Spinoza, Baruch 2

INDEX

spiritualism 129–34
spirituality, meaning of 1–3
stars 24
statistics 136, 137–50
status of science 38–9
Stevenson, Robert Louis 133
Stewart, Balfour 130, 184–5
Stokes, George Gabriel 128, 131
Strasbourg cathedral clock 21, 82
Strickland, Hugh 80, 161–2
Sumner, John Bird 63
Sun, age of 105
Sunday Lecture Society 119
Sussex, Duke of 43, 50, 155
Swainson, William 76, 77–80, 192
Synthetic Society 127

Tahiti 16
Tait, Peter Guthrie 130, 184–5
taxonomy 79–80, 161–2
teams of scientists 188
Teilhard de Chardin, Pierre 186–7
telepathy 132–3
telescope 22–3
Temple, Frederick 74, 135
Temple, William 127
Templeton Foundation 191
Tennyson, Alfred 82–3, 85, 88, 103–4, 116, 148–9, 173–4
Tenzing, Sherpa 2
Teresa, St 2
thermodynamics 105, 143
Thompson, J.V. 104
Thomson, J.J. 128, 171, 193
Thomson, William (Lord Kelvin) 101, 105, 114, 127–8, 135, 160, 171
Thoreau, Henry 168
time, tracts of 64–5
Time Machine, The (Wells) 106
Toland, John 29
Tractarians (Oxford Movement) 5, 47, 61–2
training 152–3
translators 177–8
trust 5
truth 195, 196
Turner, Frank 32
Turner, Joseph Mallord 44
two cultures 6–7
Tyndall, John 88–91, 108, 114, 125, 151, 171, 173, 191, 193

uncertainty 24–5

Unitarianism 10, 12, 14–15, 39, 61, 131, 145, 162, 193
universities 58–60, 74; *see also under individual universities*
University of London 71
Ussher, James 61

value-free science 185–6
Van Mildert, Bishop 75
Vancouver, George 168
Vestiges 100–4
Victoria, Queen 121
volcanoes 19
Volta, Alessandro 169
Voltaire 8, 19
votes 62

Wallace, A.R. 79, 187, 193
war 8, 9, 84, 114, 181, 184, 188
watches 30, 82
Watson, Richard 76
Watt, James 12
weather 44, 139
Wedgwood, Josiah 12, 154
Wedgwood, Thomas 109
Wells, H.G. 106, 111, 127
Wesley, Charles 14
Wesley, John 14, 39, 61, 108, 152
Whately, Richard 113
Whewell, William 44, 47–9, 104, 110, 140, 158, 192
Whiston, William 39
White, Gilbert 39, 155
Wight, George 68
Wilberforce, Samuel 39, 71, 116, 157, 164–6
Wilson, David 191
Wilton Diptych 194
Withers, Mrs 177
Wollaston, William 24–5
Wollaston, William Hyde 153, 164, 189
women 104, 173–80, 182, 185
wonder 86, 186–8
Wordsworth, Charles 156
Wordsworth, William 14, 82, 83, 85
working men 117–18
Wren, Christopher 142
Wurtz, Adolphe 164

X-Club 127

Yorkshire Philosophical Society museum 159
Young, George 63, 153, 159, 192

Spirituality not the same as religion (p 2) but should all these personalities be included? eg Banks (pp 16-18)

Witticisms not always clear - 14, 16